Applied Electromagnetics

J. E. Parton

S. J. T. Owen
Oregon State University

M.S. Raven
University of Nottingham

Applied
Electromagnetics

Second Edition

Springer-Verlag New York Inc.

First published 1986 by
MACMILLAN EDUCATION LTD
London and Basingstoke

Sole distributors in the USA and its dependencies
Springer-Verlag New York Inc.
175 Fifth Avenue,
New York, NY 10010
USA.

Printed in Hong Kong

ISBN 0–387–91279–7 Springer-Verlag New York

Library of Congress Cataloging in Publication Data
Parton, J. E. (John Edwin)
    Applied electromagnetics.

    Bibliography: p.
    Includes index.
    1. Electromagnetic theory. I. Owen, S. J. T.
(Sidney John Thomas) II. Raven, M. S. III. Title.
QC670.P37   1986   530.1$'$41   86–13039
ISBN 0–387–91279–7

Contents

CONTENTS ix

Preface to the First Edition

Electromagnetic theory has been a basic subject taught for more than a century to physics students but not to the electrical-engineering student. Before the Second World War the engineer was well grounded in circuit theory but was notoriously weak in field theory; by and large he might have heard of Maxwell's equations but he certainly did not use them. Since the Second World War, many factors have greatly changed the engineer's outlook; particularly the astonishing advances in electronics, in communications (particularly microwaves) and more recently in solid-state devices. Consequently, a basic course in electromagnetics and applications has been included in most first-degree courses in electrical and electronic engineering since about 1950.

The many earlier excellent texts available were unsuitable for engineering courses in electromagnetics for two reasons. First, they had been written from the point of view of the physicist, being more concerned with basic principles than with applications. Second, the introduction of SI (rationalised MKS) units meant that these earlier texts needed to be revised. Consequently the new texts in this subject have been in the main written by and for electrical engineers: as examples see the books by Skilling, Cullwick, Carter, Hayt, and Lorrain and Corson. These excellent texts have been found too advanced and too lengthy for the short time allocated to electromagnetism at Nottingham, that is about fifteen lecture hours in the first year and about twenty in the second year. In the final year there are optional subjects which do develop specialist aspects of electromagnetism such as microwaves and radiation.

It has therefore been convenient to develop this short, compact text which covers the basic material of the first two years without attempting to develop specialist needs that are already well provided. Each author has conducted the second-year course in electromagnetics for about five years, so there has been considerable modification in the choice and presentation of subject matter as a result of feedback from the students. Chapter 10, on low-frequency applications, has been based on notes from Professor L. R. Blake, who has given one or two lectures in the second-year course for several sessions.

Each chapter is provided with a set of about ten problems, mostly of a quantitative nature. Some are provided with answers; the others are deliberately left without. A student can thus be assigned a suitable 'mix' by the tutor. The examples are mostly drawn from the degree examinations at Nottingham. Appendix 2 lists the information made available to students during these examinations to avoid unnecessary memorising.

It is hoped that the text will prove acceptable to first-degree engineering students at universities and polytechnics. We shall welcome information about errors and suggestions for future modifications.

Nottingham   J. E. PARTON
July 1974   S. J. T. OWEN

Preface to the Second Edition

Since the first edition of this book was published, fundamental advances in Solid State Electronics have lead to rapid changes in Electronic Engineering and Computer Technology. Nevertheless, the fundamentals of Electromagnetism have not changed. Electromagnetic theory is consistent with the theory of relativity and to quote Einstein "The special theory of relativity has crystallized out from the Maxwell–Lorentz theory of electromagnetic phenomena" . . . "which in no way opposes the theory of relativity." Consequently, physicists and electrical and electronic engineers continue to apply Maxwell's equations to a diverse range of problems. However, electromagnetic theory deals with macroscopic phenomena; it does not consider the atomic nature of materials. A similar situation arises in classical mechanics and thermodynamics. The basic microscopic physics is implicit and care must be taken to use the correct physics for the problem concerned. For example, the field equation for Ohm's law, $J = \sigma E$, applies only under limited conditions and usually J is some complex function of E which requires a knowledge of quantum mechanics.

Although large main frame computers have been used to solve electromagnetic field problems for many years, the present widespread availability of small computers allows individual students to interact with a machine. For this reason we have incorporated additional material in chapter 8 on the numerical solution of field problems. The programming language BASIC has been used for the worked solutions since this language is available on nearly all small computers. BASIC varies slightly between different computers but it is a simple matter to modify the programs given to suit another computer.

The use of a computer can be helpful in illustrating difficult three-dimensional concepts as well as solving problems numerically. However, it is *even more important to understand theoretical concepts* in order to set up computer solutions. In addition, analytical solutions to problems are nearly always sought, if available, since these frequently reveal general principles which may not be apparent from numerical solutions. If analytical solutions are not obtainable, field sketches may provide simple and rapid approximations for checking solutions. In any case, the availability of more than one independent method of solving a field problem is considered necessary.

Electromagnetic field theory is also a theory of optics which is rapidly developing, particularly in the field of communications. In this second edition we have included new sections on an introduction to *fibre optics*. This follows naturally from High-frequency Effects and from Fresnel's equations in chapter 11.

A number of worked solutions have been included at the end of the book, selected from problems which appear at the end of each chapter. These have been avidly tackled by Nottingham undergraduates for nearly a decade.

Finally, the opportunity has been taken to make a few minor corrections to the first edition.

Nottingham   M.S. RAVEN
December, 1984

1

Vector Analysis

The application of electromagnetism in situations of concern to the scientist and technologist requires a basic knowledge of vector analysis. Vector analysis is quite properly regarded as a subject in its own right and has many other applications such as in applied mechanics and crystallography as well as its very important application in electromagnetism. It is necessary, however, to discuss the basic principles of vector analysis and to introduce the readers to the recognised notation before considering its use in describing the behaviour of electromagnetic fields and waves. To begin with the definition of *scalar* and *vector* quantities are given.

A *scalar* is a physical quantity which is completely defined by its magnitude. Examples include mass, time, temperature, current and voltage.

A *vector* is a physical quantity which is not completely defined unless its direction as well as its magnitude is specified. Examples of this include displacement, velocity, force, electric field, current density and magnetic field.

In books and printed papers vectors are printed in boldface type while scalars and the magnitude of vectors are printed in normal type. It is usual to indicate vectors in written notes by a bar, either above or below the quantity.

1.1 Vector addition and subtraction

A vector may be represented by a straight line with an arrowhead pointing in the direction of the vector and of length scaled to represent its magnitude.

The sum of two vectors is obtained by the parallelogram rule; the difference is obtained similarly, after first reversing the vector with the negative sign, see figure 1.1.

A vector F may clearly be resolved into components

$$F = a_x F_x + a_y F_y + a_z F_z$$

1

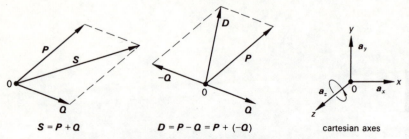

Figure 1.1   *Illustration of the sum and difference of two vectors and the cartesian right-hand-screw coordinate system.*

In this the cartesian right-hand-screw system is used though there are many other sets of axes mutually orthogonal; a_x, a_y, a_z are unit vectors parallel to x, y, z axes respectively.

Thus if

$$P = a_x P_x + a_y P_y + a_z P_z$$

and

$$Q = a_x Q_x + a_y Q_y + a_z Q_z$$

then

$$S = P + Q = a_x(P_x + Q_x) + a_y(P_y + Q_y) + a_z(P_z + Q_z)$$

and

$$D = P - Q = a_x(P_x - Q_x) + a_y(P_y - Q_y) + a_z(P_z - Q_z)$$

In many texts the cartesian unit vectors are i, j and k; here a_x, a_y and a_z are used to avoid confusion with j used as the complex-number operator $\sqrt{-1}$.

1.2   Multiplication with vectors

This is not as easy as with scalars alone and has several forms.

1.2.1   Simple product mP

This merely changes the length of P by the scalar factor m. When the scalar has a negative sign (for example $-m$) the direction is reversed. Multiplication by a scalar is shown in figure 1.2.

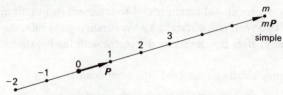

Figure 1.2   *Multiplication of a vector P by a scalar m.*

1.2.2   Scalar double product P . Q

This is a scalar quantity numerically equal to the product of the magnitudes of the vectors and the cosine of the angle between them $P . Q = PQ \cos \theta = Q . P$. It is often called a *dot* product and is illustrated in figure 1.3.

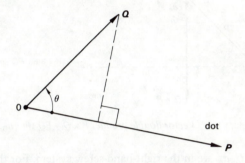

Figure 1.3   *Scalar double product P . Q = PQ cos θ.*

The distributive law holds $P . (Q + R + \ldots) = P . Q + P . R + \ldots$. In particular for the cartesian unit vectors a_x, a_y, a_z

$$a_x . a_x = a_y . a_y = a_z . a_z = +1$$

$$a_x . a_y = a_y . a_z = a_z . a_y = 0$$

If

$$P = a_x P_x + a_y P_y + a_z P_z \qquad \text{and} \qquad Q = a_x Q_x + a_y Q_y + a_z Q_z$$

then

$$P . Q = P_x Q_x + P_y Q_y + P_z Q_z$$

A typical example is the work done by a force P in moving its point of application through distance Q, which is $PQ \cos \theta$.

1.2.3   Vector double product P × Q

This is a vector of magnitude $PQ \sin \theta$ which is normal to the plane containing P and Q and its direction is given by the right-hand-screw system, see figure 1.4.

In this a_n is a unit vector normal to the P, Q plane and $-a_n$ is a unit vector normal to the Q, P plane.

$$P \times Q = PQ \sin \theta a_n$$

$$Q \times P = -PQ \sin \theta a_n = PQ \sin \theta (-a_n)$$

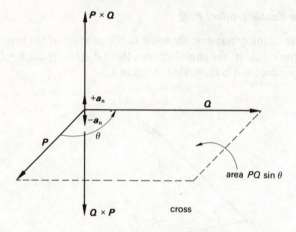

Figure 1.4   *Vector double product $P \times Q = PQ \sin \theta\, a_n$.*

Note again that P, Q, a_n are in the right-hand-screw system. For the cartesian unit vectors a_x, a_y, a_z

$$a_x \times a_x = a_y \times a_y = a_z \times a_z = 0; \qquad\qquad \theta = 0; \qquad \text{self}$$

$$a_x \times a_y = +a_z; \qquad a_y \times a_z = +a_x; \qquad a_z \times a_x = +a_y; \qquad \theta = \pi/2; \qquad \text{cyclic}$$

$$a_y \times a_x = -a_z; \qquad a_z \times a_y = -a_x; \qquad a_x \times a_z = -a_y; \qquad \theta = \pi/2; \qquad \text{anti-cyclic}$$

When P and Q are written in cartesians, the cross product is elegantly shown in determinant form.

$$P \times Q = (a_x P_x + a_y P_y + a_z P_z) \times (a_x Q_x + a_y Q_y + a_z Q_z)$$

$$= a_x(P_y Q_z - P_z Q_y) + a_y(P_z Q_x - P_x Q_z) + a_z(P_x Q_y - P_y Q_x)$$

$$= \begin{vmatrix} a_x & a_y & a_z \\ P_x & P_y & P_z \\ Q_x & Q_y & Q_z \end{vmatrix}$$

A typical example is the mechanical force dF experienced by an element of wire

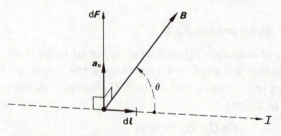

Figure 1.5   *Mechanical force dF on a current element $I\,dl$ in a magnetic field of flux density B.*

of length dl when carrying a current of magnitude I in a magnetic field of density B

$$dF = I \, dl \times B = I \, dl \, B \sin \theta a_n$$

as shown in figure 1.5.

There are three meaningful triple products of the vectors P, Q, R. They are $(P \cdot Q)R$; $P \cdot (Q \times R)$; $P \times (Q \times R)$.

1.2.4 Simple triple product $(P \cdot Q)R$

Here the dot product gives a scalar, hence the vector R has its magnitude, see figure 1.6, increased by $m = P \cdot Q = PQ \cos \theta$. Clearly $(P \cdot Q)R \neq P(Q \cdot R)$, etc.

Figure 1.6   Simple triple product $(P \cdot Q)R = mR$.

1.2.5 Scalar triple product $P \cdot (Q \times R)$

For convenience this is written $|PQR|$ and is the volume of the parallelepiped formed by P, Q, R as sides in figure 1.7.

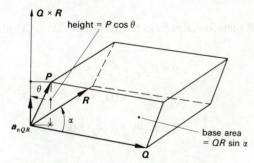

Figure 1.7   Scalar triple product $P \cdot (Q \times R)$.

$$Q \times R = (\text{base area})a_{n(QR)}$$

$$P \cdot (Q \times R) = (P \cos \theta)(\text{base area}) = (\text{height})(\text{base}) = \text{volume}$$

As long as the cyclic order P, Q, R is maintained, the position of the dot and cross is immaterial. The sign changes if the cyclic order is disturbed. Again a determinant display can be used for the scalar triple product

$$|PQR| = P \cdot (Q \times R) = -(P \times R) \cdot Q = \begin{vmatrix} P_x & P_y & P_z \\ Q_x & Q_y & Q_z \\ R_x & R_y & R_z \end{vmatrix}$$
$$= Q \cdot (R \times P) = -(Q \times P) \cdot R$$
$$= R \cdot (P \times Q) = -(R \times Q) \cdot P$$

1.2.6   Vector triple product $P \times (Q \times R)$

The result cannot depend upon the choice of axes, so make them convenient as in figure 1.8. Q is on the x axis and R lies in the xy plane. Then $(Q \times R)$ will be a vector lying on the z axis. $P \times (Q \times R)$ will be a vector perpendicular to both P and

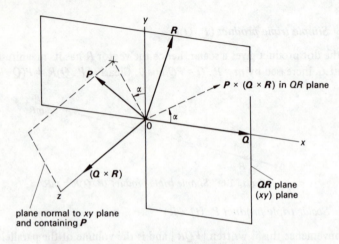

Figure 1.8   *Vector triple product $P \times (Q \times R)$.*

$(Q \times R)$ and will therefore also lie in the xy plane (that is QR plane). With this choice we have

$$P = P_x a_x + P_y a_y + P_z a_z$$

$$Q = Q_x a_x + 0a_y + 0a_z$$

$$R = R_x a_x + R_y a_y + 0a_z$$

from which

$$P \cdot Q = P_x Q_x$$

and

$$P \cdot R = P_x R_x + P_y R_y$$

Now

$$(Q \times R) = \begin{vmatrix} a_x & a_y & a_z \\ Q_x & 0 & 0 \\ R_x & R_y & 0 \end{vmatrix} = Q_x R_y a_z$$

that is along the z axis. From this we can write for the vector triple product

$$P \times (Q \times R) = \begin{vmatrix} a_x & a_y & a_z \\ P_x & P_y & P_z \\ 0 & 0 & Q_xR_y \end{vmatrix}$$

$$= P_yQ_xR_ya_x - P_xQ_xR_ya_y$$

$$= Q_xa_x(P_yR_y + P_xR_x) - P_xQ_x(a_xR_x + a_yR_y)$$

that is $P \times (Q \times R) = |P, Q, R| = Q(P \cdot R) - R(P \cdot Q)$

Note that

$$(P \times Q) \times R = -R \times (P \times Q) = Q(R \cdot P) - P(R \cdot Q) \neq P \times (Q \times R)$$

Thus the cyclic order is important and it is essential to take the bracket cross product $(Q \times R)$ first.

1.3   Vector differentiation

A variable vector has its magnitude or its direction or both varying, otherwise it is constant. Let the vector P change by $\mathrm{d}P$ in time $\mathrm{d}t$, as shown in figure 1.9. The rate of change of P, with respect to time t, $\mathrm{d}P/\mathrm{d}t$, is equal to the limit of $\delta P/\delta t$ as $\delta t \to 0$.

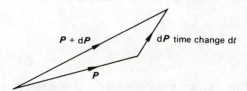

Figure 1.9   *Vector differentiation in time.*

1.3.1   *Single vector*

If $P = P_xa_x + P_ya_y + P_za_z$ in which P_x, P_y, P_z are all functions of time, then

$$\frac{\mathrm{d}P}{\mathrm{d}t} = \frac{\mathrm{d}P_x}{\mathrm{d}t}a_x + \frac{\mathrm{d}P_y}{\mathrm{d}t}a_y + \frac{\mathrm{d}P_z}{\mathrm{d}t}a_z$$

1.3.2   *Vector sum*

If P and Q are both functions of time t, then

$$\frac{\mathrm{d}}{\mathrm{d}t}(P + Q) = \frac{\mathrm{d}P}{\mathrm{d}t} + \frac{\mathrm{d}Q}{\mathrm{d}t}$$

1.3.3 Dot product

$P . Q$ = scalar S and P, Q and S all vary with time.

$$S + \mathrm{d}S = (P + \mathrm{d}P) . (Q + \mathrm{d}Q) = P . Q + \mathrm{d}P . Q + P . \mathrm{d}Q + \mathrm{d}P . \mathrm{d}Q$$

Hence

$$\mathrm{d}S = \mathrm{d}P . Q + P . \mathrm{d}Q + \mathrm{d}P . \mathrm{d}Q$$

(second order negligible) and

$$\frac{\mathrm{d}S}{\mathrm{d}t} = \frac{\mathrm{d}}{\mathrm{d}t} (P . Q) = P . \frac{\mathrm{d}Q}{\mathrm{d}t} + Q . \frac{\mathrm{d}P}{\mathrm{d}t}$$

1.3.4 Cross product

$P \times Q$ = vector R and P, Q, and R are functions of time.

$$R + \mathrm{d}R = (P + \mathrm{d}P) \times (Q + \mathrm{d}Q) = P \times Q + \mathrm{d}P \times Q + P \times \mathrm{d}Q + \mathrm{d}P \times \mathrm{d}Q$$

and

$$\mathrm{d}R = P \times \mathrm{d}Q + \mathrm{d}P \times Q + (\mathrm{d}P \times \mathrm{d}Q)$$

(second order negligible)

$$\frac{\mathrm{d}R}{\mathrm{d}t} = \frac{\mathrm{d}}{\mathrm{d}t} (P \times Q) = P \times \frac{\mathrm{d}Q}{\mathrm{d}t} + \frac{\mathrm{d}P}{\mathrm{d}t} \times Q$$

Note that the order P before Q must be maintained.

1.4 Fields

The values of certain physical quantities vary with position in space and such a quantity may be expressed as a continuous function of the space coordinates. It may depend upon one or more of the coordinates x, y, z. The region throughout which a physical quantity is so specified is a *field*. A field may be a scalar or a vector.

1.4.1 Scalar field S

In such a field the quantity may be represented by a continuous function, say $S(x, y, z)$, giving its value at all points (x, y, z). In practice scalar functions have no abrupt changes and they can be 'mapped' by constructing a series of surfaces on each of which the scalar quantity has the same value. These surfaces are referred to as 'levels' and no two surfaces may meet. Scalar fields include temperature distribution and electric potential with corresponding levels of temperature isotherms and electric equipotentials.

1.4.2 Vector field F

In a vector field the physical quantity is represented by a continuous vector function, say $F(x, y, z)$, which specifies the magnitude and direction of F at each point (x, y, z). A vector field is 'mapped' by constructing lines of flux that are everywhere tangential to F; they are sometimes called flow lines or vector lines. The closeness or density of these lines can be used as a measure of the magnitude of the field. Examples of vector fields include fluid velocity and electric force on a test charge; the corresponding lines of flux are flow lines in a fluid and flux lines in an electric field.

1.5 Gradient, divergence and curl

In describing physical quantities certain particular functions of the space derivatives of scalar and vector quantities are of great use. One of these concerns an operation on a scalar field and two of these are associated with operations on vector fields. These functions will now be defined.

The gradient of a scalar is a vector function whose magnitude at any point in the scalar field is the maximum rate of change of the scalar function, with respect to distance, at that point. The direction along which the maximum rate of change occurs is the direction of the vector. The gradient of a scalar field S is written grad S and

$$\text{grad } S = \frac{\partial S}{\partial a_n} a_n$$

where a_n is a unit normal vector at the point so chosen to maximise the derivative.

The divergence of a vector is defined as the total outward flux of a vector per unit volume as the volume shrinks to a point. The flux of a vector F through an infinitesimal surface dS bounding a volume Δvol is $F \cdot dS$, where dS is directed normal to the surface and represents the elemental surface area. Thus the divergence of a vector, written div F, may be defined by

$$\text{div } F = \lim_{\Delta\text{vol} \to 0} \frac{\oint_{\Delta S} F \cdot dS}{\Delta\text{vol}}$$

Although $\oint_{\Delta S} F \cdot dS$ may tend to zero as Δvol tends to zero the value of div F can approach a non-zero value, since in effect the volume occupied by a point is zero. The divergence of a vector gives the amount of outward flux and is therefore a scalar point function.

The line integral around a closed path l in a vector field F, $\oint_l F \cdot dl$, is called the circulation of F. Circulation is a term often used in fluid mechanics and aerodynamics where F would represent the velocity of a fluid. The circulation of F is obtained by multiplying the component of F parallel to the path by the differential path length, and then carrying out a summation over the whole path length as the differential path length tends to zero. The curl of a vector F is defined by the relationship

$$\text{curl } F = \lim_{\Delta S \to 0} \frac{\oint_l F \cdot dl}{\Delta S} a_n$$

where ΔS is an infinitesimal area in the vector field, the periphery of which is the path l, and a_n is a unit normal vector, the direction of which maximises the circulation. The closed-loop integral is carried out around the periphery of the infinitesimal area ΔS. It is possible to regard the normal component of curl F physically to be the limit of the circulation per unit area as that area shrinks to a point. The curl of a vector is another vector and, although $\oint_l F \cdot dl$ can tend to zero as ΔS tends to zero, curl F may still have a non-zero value.

In these definitions the symbols $\oint_{\Delta S}$ and $\oint_l$ indicate that the integrations are carried out for closed surfaces and closed paths respectively.

It is often useful in manipulations involving vector operations to use the vector operator del, written ∇. This was first introduced by W. R. Hamilton, and O. Heaviside called it 'nabla' after the Assyrian word for a harp. J. C. Maxwell used the name 'atled', the reverse of delta and the first use of del, now universally adopted, was by J. W. Gibbs. In rectangular coordinates

$$\text{del} = \nabla = a_x \frac{\partial}{\partial x} + a_y \frac{\partial}{\partial y} + a_z \frac{\partial}{\partial z}$$

∇ is not a vector but is a differential vector operator and cannot be interpreted without an operand. It is defined only in the rectangular cartesian coordinate system.

1.5.1 The gradient of a scalar

Suppose S to be the temperature in a body, mapped out by 'level' surfaces, as shown in figure 1.10, on which S is a constant.

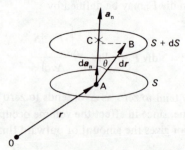

Figure 1.10   *Gradient of a scalar* ∇S.

$$OB = r + dr$$

a_n is a unit normal vector.

$$da_n = dr \cos \theta$$

The rate of increase of S in direction AB = $\partial S/\partial r = (\partial S/\partial a_n) \cos \theta$. This will be greatest if direction AC is chosen when

$$\text{greatest rate of increase of } S = \frac{\partial S}{\partial a_n} a_n = \text{grad } S$$

$$\text{grad } S = \frac{\partial S}{\partial x} a_x + \frac{\partial S}{\partial y} a_y + \frac{\partial S}{\partial z} a_z$$

It is possible to relate the operation of ∇ on a scalar field function S to the gradient as follows. Consider the dot product

$$(\text{grad } S) \cdot dr = \frac{\partial S}{\partial a_n} a_n \cdot dr = \frac{\partial S}{\partial a_n} dr \cos \theta = \frac{\partial S}{\partial a_n} da_n = dS$$

$$= \frac{\partial S}{\partial x} dx + \frac{\partial S}{\partial y} dy + \frac{\partial S}{\partial z} dz$$

$$= \left(\frac{\partial S}{\partial x} a_x + \frac{\partial S}{\partial y} a_y + \frac{\partial S}{\partial z} a_z \right) \cdot (dx\, a_x + dy\, a_y + dz\, a_z)$$

$$= \nabla S \cdot dr$$

The operations grad and del applied to a scalar S are identical and a vector field is produced which is said to be lamellar. This term is used because such fields may have their space mapped out, by equipotential surfaces, into thin sheets (lamellae). An essential property of such a derived vector field is that the line integral of F around any closed path in the field is necessarily zero. Any field satisfying this property, namely $\oint F \cdot dl = 0$, is named a *conservative field*. The static electric field is an example of this type of field. The work involved in moving a test charge q around a closed loop in a static electric field of electric field strength E

$$\oint_l qE \cdot dl = 0$$

is a statement of the conservation of energy. Another field exhibiting this property is the *gravitational field* but it is *not true* of the *time-varying electric field*.

1.5.2 The divergence of a vector

Consider an elementary cube $dx\, dy\, dz$ in the vector field F shown in figure 1.11. F_x, F_y, F_z exist at the middle of the dx, dy, dz cube, and the components of F at the various faces of the cube are given by

$$(F_{x^+}) = F_x + \frac{1}{2} \frac{\partial F_x}{\partial x} dx \qquad (F_{x^-}) = F_x - \frac{1}{2} \frac{\partial F_x}{\partial x} dx$$

$$(F_{y^+}) = F_y + \frac{1}{2} \frac{\partial F_y}{\partial y} dy \qquad (F_{y^-}) = F_y - \frac{1}{2} \frac{\partial F_y}{\partial y} dy$$

$$(F_z{}^+) = F_z + \frac{1}{2}\frac{\partial F_z}{\partial z}\,dz \qquad (F_z{}^-) = F_z - \frac{1}{2}\frac{\partial F_z}{\partial z}\,dz$$

The outward flux of the vector may be calculated by considering the outgoing flux through each face of the elementary cube. The flux through the dy dz face is, in the positive x direction

$$\left(F_x + \frac{1}{2}\frac{\partial F_x}{\partial x}\,dx\right)dy\,dz$$

and that through the opposite face is

$$\left(F_x - \frac{1}{2}\frac{\partial F_x}{\partial x}\,dx\right)dy\,dz$$

The net outward flux through these two faces is

$$(F_x{}^+)\,dy\,dz - (F_x{}^-)\,dy\,dz = \frac{\partial F_x}{\partial x}\,dx\,dy\,dz$$

Similarly the net outward flux through the dx dy faces is

$$(F_z{}^+)\,dx\,dy - (F_z{}^-)\,dx\,dy = \frac{\partial F_z}{\partial z}\,dx\,dy\,dz$$

and through the dx dz faces is

$$(F_y{}^+)\,dz\,dx - (F_y{}^-)\,dz\,dx = \frac{\partial F_y}{\partial y}\,dx\,dy\,dz$$

The net flux emerging from dx dy dz is given by

$$\left(\frac{\partial F_x}{\partial x} + \frac{\partial F_y}{\partial y} + \frac{\partial F_z}{\partial z}\right)dx\,dy\,dz$$

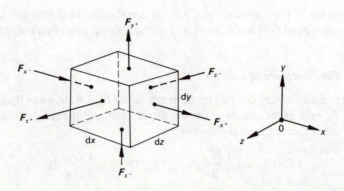

Figure 1.11   *Divergence of a vector* $\nabla.F$.

and the divergence of F is given by

$$\text{div } F = \frac{\text{net flux}}{\text{volume}} = \frac{\partial F_x}{\partial x} + \frac{\partial F_y}{\partial y} + \frac{\partial F_z}{\partial z}$$

The divergence of a vector F may be shown to be equal to the scalar product of ∇ and F

$$\nabla \cdot F = \left(\frac{\partial}{\partial x} a_x + \frac{\partial}{\partial y} a_y + \frac{\partial}{\partial y} a_z \right) \cdot (F_x a_x + F_y a_y + F_z a_z)$$

$$= \frac{\partial F_x}{\partial x} + \frac{\partial F_y}{\partial y} + \frac{\partial F_z}{\partial z} = \text{div } F = \text{scalar } S$$

Clearly in a field where there is neither source nor sink of flux, div $F = 0$. Such a field is said to be *solenoidal*; the lines of flux must form closed curves or must terminate on a bounding surface or must extend to infinity. An example would be flow of an incompressible fluid (that is its density is not changing). When div F is positive it must be at a *source* in the field; conversely when div F is negative it must be at a *sink*. An example of a solenoidal vector field is the current density J associated with static fields. The current flow governed by static fields flows in closed paths because clearly there can be no accumulation of charge at a point. However, div $J = 0$ does not hold in all cases for time-varying fields.

1.5.3 The curl of a vector

In order to illustrate the concept of the curl of a vector consider the closed rectangular path ABCD in the yz plane, as shown in figure 1.12 in which AB = DC = dy, and BC = AD = dz. The values of the components of the vector at each line of the

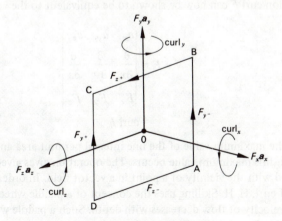

Figure 1.12 *Curl of a vector* $\nabla \times F$.

rectangular path are

$$\text{AB} \qquad F_y - \frac{1}{2}\frac{\partial F_y}{\partial z}\, dz = (F_y\text{-})$$

$$\text{BC} \qquad F_z + \frac{1}{2}\frac{\partial F_z}{\partial y}\, dy = (F_z\text{+})$$

$$\text{DC} \qquad F_y + \frac{1}{2}\frac{\partial F_y}{\partial z}\, dz = (F_y\text{+})$$

$$\text{AD} \qquad F_z - \frac{1}{2}\frac{\partial F_z}{\partial y}\, dy = (F_z\text{-})$$

The line integral of $F . dl$ around ABCD is given by

$$\oint_{\text{ABCD}} F . \, dl = \text{AB}(F_y\text{-}) + \text{BC}(F_z\text{+}) - \text{CD}(F_y\text{+}) - \text{DA}(F_z\text{-})$$

$$= dy\left(-\frac{\partial F_y}{\partial z}\, dz\right) + dz\left(+\frac{\partial F_z}{\partial y}\, dy\right)$$

From the previous definition of curl F the x component of the curl of the vector field is given by

$$\text{curl}_x = \frac{\displaystyle\oint_{\text{ABCD}} F . \, dl}{\text{area ABCD}} = \left(\frac{\partial F_z}{\partial y} - \frac{\partial F_y}{\partial z}\right)a_x$$

The components curl_y and curl_z may be found by taking search areas $dx\, dz$ and $dx\, dy$ respectively. Hence

$$\text{curl } F = \left(\frac{\partial F_z}{\partial y} - \frac{\partial F_y}{\partial z}\right)a_x + \left(\frac{\partial F_x}{\partial z} - \frac{\partial F_z}{\partial x}\right)a_y + \left(\frac{\partial F_y}{\partial x} - \frac{\partial F_x}{\partial y}\right)a_z$$

The operation curl F can now be shown to be equivalent to the vector product of ∇ and F

$$\nabla \times F = \begin{vmatrix} a_x & a_y & a_z \\ \dfrac{\partial}{\partial x} & \dfrac{\partial}{\partial y} & \dfrac{\partial}{\partial z} \\ F_x & F_y & F_z \end{vmatrix}$$

$$= \text{curl } F$$

This gives the maximum value of the line integral per unit area and also the direction along which this maximum value occurs. The operation gives a vector, curl F, which is concerned with the vicinity of a point in a vector field. In order to illustrate the principle of curl, H. H. Skilling uses the concept of a paddle wheel in a river, in which the velocity of flow decreases with depth. Such a paddle wheel would rotate in a clockwise direction, see figure 1.13, and its axis when rotating with maximum velocity would give the direction of curl. Here F has not been derived from a scalar

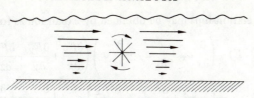

Figure 1.13   *Illustration of curl.*

field S by means of its gradient or else the line integral around a closed path in the corresponding vector field would have been zero. Although curl suggests that motion or flow in curves is present this is not strictly necessary, as in the example quoted.

A field in which the curl is zero is generally called *irrotational* and by definition is also *lamellar* since $\oint F \cdot dl = 0$. In general a conservative field is both irrotational and lamellar but by common usage the former term has now almost entirely replaced the latter. In the paddle-wheel example such a field would not cause rotation.

This section has introduced the del operator ∇ but it has been defined only in terms of the rectangular cartesian coordinate system. Therefore the equivalence of ∇S and gradient of S, of $\nabla \cdot F$ and the divergence of F and of $\nabla \times F$ and the curl of F is strictly true only for the cartesian coordinate system and in other coordinate systems the expressions are more complex. The discussion of other coordinate systems and the vector identities in those systems are included in appendixes 1 and 2.

1.6   Double operations with del

The operator ∇ has been used to carry out the three distinct physical operations of grad, div and curl. More complex operations may be expressed in terms of these three basic ones. Taking pairs of operations in sequence, and remembering the restrictions

$$\text{grad}(\text{scalar } S) = (\text{vector } P)$$

$$\text{div}(\text{vector } F) = (\text{scalar } T)$$

$$\text{curl}(\text{vector } F) = (\text{vector } Q)$$

the double operations may be represented in table form (table 1.1).

Table 1.1

	grad S	div F	curl F
grad	no	yes	no
div	yes	no	zero
curl	zero	no	yes

The allowed relationships may be obtained by expansion as follows

$$\text{grad div } F = \nabla(\nabla . F) = \left(\frac{\partial}{\partial x} a_x + \frac{\partial}{\partial y} a_y + \frac{\partial}{\partial z} a_z\right)\left(\frac{\partial F_x}{\partial x} + \frac{\partial F_y}{\partial y} + \frac{\partial F_z}{\partial z}\right)$$

$$= \left(\frac{\partial^2 F_x}{\partial x^2} + \frac{\partial^2 F_y}{\partial x\,\partial y} + \frac{\partial^2 F_z}{\partial x\,\partial z}\right) a_x + \left(\frac{\partial^2 F_x}{\partial x\,\partial y} + \frac{\partial^2 F_y}{\partial y^2} + \frac{\partial^2 F_z}{\partial y\,\partial z}\right) a_y$$

$$+ \left(\frac{\partial^2 F_x}{\partial x\,\partial z} + \frac{\partial^2 F_y}{\partial y\,\partial z} + \frac{\partial^2 F_z}{\partial z^2}\right) a_z$$

$$\text{div grad } S = \nabla . \nabla S = \left(\frac{\partial}{\partial x} a_x + \frac{\partial}{\partial y} a_y + \frac{\partial}{\partial z} a_z\right) . \left(\frac{\partial S}{\partial x} a_x + \frac{\partial S}{\partial y} a_y + \frac{\partial S}{\partial z} a_z\right)$$

$$= \left(\frac{\partial}{\partial x^2} + \frac{\partial}{\partial y^2} + \frac{\partial}{\partial z^2}\right) S = \nabla^2 S$$

The operator ∇^2 is known as the *laplacian or the Laplace operator* and is written as *lap*.

$$\text{div curl } F = \nabla . \nabla \times F$$

$$= \nabla . \begin{vmatrix} a_x & a_y & a_z \\ \dfrac{\partial}{\partial x} & \dfrac{\partial}{\partial y} & \dfrac{\partial}{\partial z} \\ F_x & F_y & F_z \end{vmatrix}$$

$$= \frac{\partial}{\partial x}\left(\frac{\partial F_z}{\partial y} - \frac{\partial F_y}{\partial z}\right) + \frac{\partial}{\partial y}\left(\frac{\partial F_x}{\partial z} - \frac{\partial F_z}{\partial x}\right) + \frac{\partial}{\partial z}\left(\frac{\partial F_y}{\partial x} - \frac{\partial F_x}{\partial y}\right)$$

$$= 0$$

$$\text{curl grad } S = \nabla \times \nabla S = \nabla \times \left(\frac{\partial S}{\partial x} a_x + \frac{\partial S}{\partial y} a_y + \frac{\partial S}{\partial z} a_z\right)$$

$$= \left(\frac{\partial^2 S}{\partial y\,\partial z} - \frac{\partial^2 S}{\partial z\,\partial y}\right) a_x + \left(\frac{\partial^2 S}{\partial z\,\partial x} - \frac{\partial^2 S}{\partial x\,\partial z}\right) a_y$$

$$+ \left(\frac{\partial^2 S}{\partial x\,\partial y} - \frac{\partial^2 S}{\partial y\,\partial x}\right) a_z$$

$$= 0$$

$$\text{curl curl } F = \nabla \times (\nabla \times F)$$

$$= \nabla \times \begin{vmatrix} a_x & a_y & a_z \\ \dfrac{\partial}{\partial x} & \dfrac{\partial}{\partial y} & \dfrac{\partial}{\partial z} \\ F_x & F_y & F_z \end{vmatrix}$$

$$= \begin{vmatrix} a_x & a_y & a_z \\ \dfrac{\partial}{\partial x} & \dfrac{\partial}{\partial y} & \dfrac{\partial}{\partial z} \\ \left(\dfrac{\partial F_z}{\partial y} - \dfrac{\partial F_y}{\partial z}\right) & \left(\dfrac{\partial F_x}{\partial z} - \dfrac{\partial F_z}{\partial x}\right) & \left(\dfrac{\partial F_y}{\partial x} - \dfrac{\partial F_x}{\partial y}\right) \end{vmatrix}$$

$$= \frac{\partial}{\partial x}\left(\frac{\partial F_x}{\partial x} + \frac{\partial F_y}{\partial x} + \frac{\partial F_z}{\partial z}\right) a_x + \frac{\partial}{\partial y}\left(\frac{\partial F_x}{\partial x} + \frac{\partial F_y}{\partial y} + \frac{\partial F_z}{\partial z}\right) a_y$$

$$+ \frac{\partial}{\partial z}\left(\frac{\partial F_x}{\partial x} + \frac{\partial F_y}{\partial y} + \frac{\partial F_z}{\partial z}\right) a_z$$

$$- \left(\frac{\partial^2}{\partial x^2} + \frac{\partial^2}{\partial y^2} + \frac{\partial^2}{\partial z^2}\right) (F_x a_x + F_y a_y + F_z a_z)$$

$$= \nabla(\nabla . F) - \nabla^2 F$$

The two allowed double operations of ∇ which yield zero, that is, curl grad S and div curl F, are examples of operations on derived vector fields. It has been noted in section 1.5.1 that the line integral of the vector (grad S) around a closed path is zero, hence it follows that curl grad S is also zero. The vector grad S represents a conservative field and curl grad S is lamellar and irrotational. Conversely if curl $F = 0$ then F may be expressed as the gradient of a scalar field. This is often used as an alternative criterion for the definition of a conservative field.

Obviously, if F has been derived from a scalar S by the gradient, then from the above div curl $F = 0$. This is, however, true for a general vector F. The components of curl F are the circulations per unit area as the area tends to a point and are oriented perpendicular to the vectors a_x, a_y, a_z. The vector curl F is associated with closed paths in a vector field, and has no sources or sinks associated with it, in which case the divergence of curl $F = 0$. Curl F is thus solenoidal and an alternative criterion for a solenoidal field is that it must be expressible as the curl of a vector. This particular relationship will be discussed in section 4.4.2 as an example of the use of Stokes's theorem. Note that $\nabla . (\nabla \times F)$ may be compared with the scalar triple product $P . (Q \times R)$ which yields zero if two of the vectors are the same.

1.6.1 The laplacian

The laplacian or lap operator in the rectangular cartesian coordinate system is $\partial^2/\partial x^2 + \partial^2/\partial y^2 + \partial^2/\partial z^2$, and is written as ∇^2. It can operate on scalar and vector fields and $\nabla^2 S$ may be defined as the divergence of the gradient of the scalar field

$$\nabla^2 S = \nabla . \nabla S = \left(\frac{\partial^2}{\partial x^2} + \frac{\partial^2}{\partial y^2} + \frac{\partial^2}{\partial z^2}\right) S$$

$\nabla^2 F$ may be defined as the gradient of the divergence of F minus the curl of the curl of F

$$\nabla^2 F = \nabla(\nabla . F) - \nabla \times \nabla \times F$$

and in the rectangular coordinate system is equal to the vector sum of the laplacian operations on the three scalar components of the field vector F. This is true only for the rectangular coordinate system and in other coordinate systems different terms arise. The development of the operations of ∇ in other coordinate systems is given in appendix 2 and the reader is advised to pay particular attention to the expansions involving grad, div, curl and the laplacian for these other systems.

1.7   Vector field classification

The operations div and curl can be used to classify fields (figure 1.14 illustrates this pictorially) as described by Hague (1951, pp. 54-6).

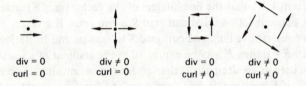

| div = 0 | div ≠ 0 | div = 0 | div ≠ 0 |
| curl = 0 | curl = 0 | curl ≠ 0 | curl ≠ 0 |

Figure 1.14   *Classification of fields by div and curl.*

1.7.1   div F = 0, curl F = 0

Since div $F = 0$ the field is solenoidal. Also because the field has no curl it must have been derived from a scalar potential field, that is $F = \text{grad } S$. Hence the field is lamellar or irrotational and we can write

$$\text{div } F = \text{div}(\text{grad } S) = \nabla . \nabla S = \nabla^2 S = 0$$

$$\frac{\partial^2 S}{\partial x^2} + \frac{\partial^2 S}{\partial y^2} + \frac{\partial^2 S}{\partial z^2} = 0 \qquad \text{Laplace equation}$$

1.7.2   div F ≠ 0, curl F = 0

The field is not solenoidal but it is irrotational or lamellar and $\nabla^2 S \neq 0$

$$\frac{\partial^2 S}{\partial x^2} + \frac{\partial^2 S}{\partial y^2} + \frac{\partial^2 S}{\partial z^2} \neq 0 \qquad \text{Poisson equation}$$

1.7.3   *div F* = 0, *curl F* ≠ 0

This is solenoidal but it is also rotational and cannot have been derived from a scalar potential S. But suppose it has been obtained from a *vector potential A* so chosen that $F = \text{curl } A = \nabla \times A$ then we have

$$\text{curl } F = \text{curl} (\text{curl } A) = \text{grad div } A - \nabla^2 A$$

We can also stipulate that the vector potential A is solenoidal so that div $A = 0$ giving finally grad div $A = 0$ and

$$\text{curl } F = -\nabla^2 A \neq 0$$

This is Poisson's equation using vector potential A instead of scalar S.

1.7.4   *div F* ≠ 0, *curl F* ≠ 0

This is the most general field, for example the rotational motion of a compressible fluid. It can be expressed as the sum of two fields. Assume that

$$F = \text{grad } S + \text{curl } A$$

$$\text{div } F = \text{div grad } S + \text{div curl } A = \nabla^2 S + 0 \neq 0 \qquad \text{Poisson in } S$$

$$\text{curl } F = \text{curl grad } S + \text{curl curl } A$$

$$= 0 + \text{grad div } A - \nabla^2 A = -\nabla^2 A \neq 0 \qquad \text{Poisson in } A$$

Hence a general field F may be replaced by *two* fields; one is lamellar and non-curl derived from the scalar potential S by grad S; the other is rotational being derived from the vector potential A by curl A.

1.8   Coordinate systems

The cartesian system has mostly been used because of its familiarity and the regularity of vector operations expressed in this system. Often it is advantageous to use other systems; appendix 1 shows how to convert from one to another of the three main systems of coordinates—cartesian, cylindrical and spherical.

1.9   Data sheet

Appendix 2 consists of data suitable for use in certain types of examination. It contains the following

(i) Fundamental physical constants.
(ii) Vector identities.
(iii) Vector operations div, grad, curl and lap in the three main coordinate systems.

Problems

The questions in this and following exercises have been drawn from a variety of sources, including departmental examinations in electrical and electronic engineering at the University of Nottingham. Where available, answers are given in parentheses ( ).

1.1 (a) Using vectors, prove that the diagonals of a parallelogram bisect each other.
(b) The position vectors of P and Q are $a_x + 3a_y - 7a_z$ and $5a_x - 2a_y + 4a_z$. Find a unit vector in the direction PQ.
$(4a_x - 5a_y + 11a_z/9\sqrt{2})$

1.2 (a) Find (i) the angle between the two vectors $4a_x + 6a_y + 2a_z$ and $2a_x - 12a_y + 2a_z$, and (ii) a unit vector perpendicular to both.
(b) Prove $(A \times B) . (C \times D) = (A . C)(B . D) - (A . D)(B . C)$.
$(130° \ 35'; \ \pm(9a_x - 1a_y - 15a_z)/\sqrt{307})$

1.3 Given $A = t^2 a_x - ta_y + (2t + 1)a_z$ and $B = (2t - 3)a_x + a_y - ta_z$,

$$\frac{d}{dt}(A . B), \qquad \frac{d}{dt}(A \times B), \qquad \frac{d}{dt}|A + B|, \qquad \text{and} \qquad \frac{d}{dt}\left(A \times \frac{dB}{dt}\right)$$

Evaluate each at time $t = 2$.

$(2, (2a_x + 24a_y + 9a_z), 34/35^{1/2}, (1a_x + 8a_y + 2a_z))$

1.4 (a) Find grad S, and div grad S for the scalar function $S_1 = 2xz^4 - x^2y$ evaluating at the point $(2, -2, 1)$.
(b) Repeat for scalar function $S_2 = r^2/\exp(r)$ evaluating at the spherical surface $r = 4$.
(c) Given vector field $F = 2x^2 za_x - xy^2 za_y + 3yz^2 a_z$, find div F, grad div F, curl F and curl curl F, evaluating these at the point $(1, 1, 1)$.
$((a) \ 10a_x - 4a_y + 16a_z, 52. \ (b) \ -0.1465a_r, -0.0366. \ (c) \ 8; 2a_x + 4a_y + 8a_z, 4a_x + 2a_y - 1a_z, -2a_x + 6a_y + 2a_z)$

1.5 Given $r = xa_x + ya_y + za_z$ and a constant vector A, evaluate (i) div r, (ii) curl r, (iii) $-$grad $(1/r)$, (iv) $\nabla^2 r^2$, (v) grad $A . r$ and (vi) curl $(A \times r)$.

$(3, 0, r/|r|^3, 6, A, 2A)$

1.6 Given a field vector $F = xa_x + 2ya_y + 3za_z$ and a scalar function $S = x^2 - y^2 + z^2$, find div (SF). Evaluate it at the points $(2, 2\sqrt{2}, 2)$ and $(1, 1, 1)$. Is SF a solenoid vector field?

$(0; 10; no)$

1.7 Classify vector fields with reference to the operations div and curl; give an example of each.

A vector field F is given by

$$F = (\tfrac{1}{3}x^3 - 2yx)a_x + (y^2 - yz)a_y + (\tfrac{1}{2}z^2 - x^2z)a_z$$

Determine to which one of your classes it belongs.

(div $F = 0$; curl $F \neq 0$)

1.8 A scalar field function S has the following properties (using cylindrical coordinates)

   (i) S is independent of z

   (ii) Laplace's equation holds for $\rho > 2$

   (iii) As $\rho \to \infty$, $S \to 0$

   (iv) On the surface $\rho = 2$, $S = \cos \phi - 3 \sin 3\phi$

For the space defined by $\rho \geqslant 2$ find the function S assuming it to be of the form $S = a\rho^m \cos \phi + b\rho^n \sin 3\phi$, or otherwise.

($S = (2/\rho) \cos \phi - (24/\rho^3) \sin 3\phi$)

1.9 The heat flow vector $Q = \kappa \nabla T$ where T is the temperature and κ the thermal conductivity. Show that where $T = z \sin x \sinh y$, then div $Q = 0$.

($\nabla^2 T = 0$)

1.10 Given $R = (a \cos \omega t)a_x + (a \sin \omega t)a_y + (b\omega t)a_z$ show that dR/dt and d^2R/dt^2 are vectors of fixed magnitudes (independent of t) and find their values.

($\omega\sqrt{(a^2 + b^2)}$; $\omega^2 a$)

2

The Electric Field

This chapter is concerned with time-independent electric fields arising from fixed or static charges and hence the term electrostatic field is often used. The starting point is usually from the experimental investigations carried out in 1785 by Cavendish and Coulomb, who likened the electrostatic field associated with the charge on a body to the gravitational field associated with the mass of a body.

2.1 Coulomb's law

As a result of experiments carried out on two isolated charged bodies whose separation was large compared with their size, Coulomb formulated a law governing the force F existing between two charged bodies. The force acts in the direction of the line joining the two bodies and has magnitude

$$F = \frac{\alpha Q q}{r^2}$$

where Q and q are the charges and r their separation as shown in figure 2.1.

Figure 2.1   *Mechanical force F between two charges Q and q.*

In SI units (Système International d'Unités), which is an extended version of the MKSA Giorgi rationalised system, $\alpha = 1/4\pi\epsilon_0$ where $\epsilon_0 \approx 10^{-9}/36\pi$ farads per metre is the permittivity of free space. The function 4π arises from spherical symmetry, in particular from the surface area $4\pi r^2$. The sign of the charges Q and q look after the direction of the force which is repulsive for like signs and attractive for unlike signs. In the approximate value of ϵ_0 it is assumed that the free space velocity of light is 3×10^8 m/s.

2.2 Electric field strength E

This is often called the electric intensity.

Consider a test charge q coulombs placed in an electric field where the field strength is E, the mechanical force $F = qE$ and $E = F/q$ newtons per coulomb or volts per metre. Here the definition precludes the test charge from having anything but a negligible effect on those charges producing the original field. Should the test charge cause a redistribution of these charges then the electric flux associated with the field would change. A more exact definition would be

$$E = \lim_{q \to 0} \frac{F}{q}$$

This definition is equivalent to defining the electric field strength as a force per unit charge. A unit charge of 1 coulomb is physically large and particularly unwieldy to use in definitions, hence the concept of a small test charge q. For the case illustrated in figure 2.2

$$E = \frac{Q}{4\pi\epsilon_0 r^2} a_r$$

F and E are in the same direction if q is positive and E is an example of a field vector.

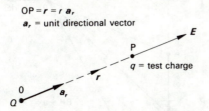

OP $= r = r\, a_r$
a_r = unit directional vector

E

P

q = test charge

r

0

a_r

Q

Figure 2.2   *Electric field strength E due to charge Q.*

2.2.1   *Field due to several charges Q_1, Q_2, Q_3, etc.*

The principle of superposition may be used to find the resultant electric field strength E due to several point charges. To obtain E due to say three concentrated charges Q_1, Q_2, Q_3 add the individual field strengths E_1, E_2, E_3 vectorially as

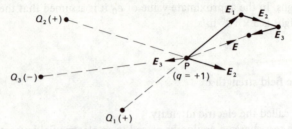

Figure 2.3   *Electric field strength due to several charges.*

shown in figure 2.3. We can clearly extend this to any number of charges m which are not necessarily coplanar.

$$E = E_1 + E_2 + \ldots + E_m = \sum_{n=1}^{m} \left(\frac{Q_n}{4\pi\epsilon_0 r_n^{\,2}} \, a_n \right)$$

in which a_n is the unit vector in direction of E_n.

2.2.2   *Field due to a line charge ρ_L coulomb/m*

This particular charge distribution is shown in figure 2.4 which is of a finite line extending 0 to A on the z axis, charged with ρ_L coulombs/m. It is customary to refer to distributed charge with the symbol ρ and a suffix L, S and V to indicate line, surface and volume charge respectively. This must not be confused with ρ

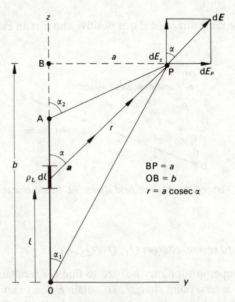

Figure 2.4   *Field due to a line charge ρ_L coulombs per metre.*

used as a coordinate in the cylindrical polar system or with the same symbol commonly used for resistivity. We wish to find the electric field strength E at $\mathbf{P}$ due to this charged line.

$$E = \int_0^A \frac{\rho_L \, \mathrm{d}l}{4\pi\epsilon_0 r^2} \, a_r$$

Suppose we consider only the normal, radial or ρ component E_ρ

$$\mathrm{d}E_\rho = \mathrm{d}E \sin\alpha = \frac{\rho_L \, \mathrm{d}l \sin\alpha}{4\pi\epsilon_0 a^2 \, \mathrm{cosec}^2 \, \alpha}$$

where, from figure 2.4

$$l = b - a \cot\alpha$$

and

$$\mathrm{d}l = +a \, \mathrm{cosec}^2 \, \alpha \, \mathrm{d}\alpha$$

Hence

$$E_\rho = \int_{\alpha_1}^{\alpha_2} \frac{\rho_L \sin\alpha \, \mathrm{d}\alpha}{4\pi\epsilon_0 a} = \frac{\rho_L}{4\pi\epsilon_0 a} (\cos\alpha_1 - \cos\alpha_2)$$

For the infinite line charge, $\alpha_1 \to 0$, $\alpha_2 \to \pi$ so that

$$E = E_\rho a_\rho = \frac{\rho_L}{2\pi\epsilon_0 a} a_\rho$$

this is the only (radial, normal) field component, for the axial field strength $E_z = 0$. The field has cylindrical symmetry.

2.2.3 Field due to a surface charge ρ_S coulombs/m^2

The electric field strength at any point P due to such a surface S may be found by integration

$$E = \int_{\text{surface}} \frac{\rho_S \, \mathrm{d}S}{4\pi\epsilon_0 r^2} \, a_r$$

where a_r is a unit vector along r, the position vector of P with respect to dS.

If we have a choice of axes these can be fixed so that the charged surface lies in the yz plane. The point P can be on the x axis, that is coordinates $(a, 0, 0)$. This is illustrated in figure 2.5a and may be used to calculate the electric field strength due to an infinitely long plate in the z direction, situated in the yz plane and symmetrical about the z axis, as shown in figure 2.5b.

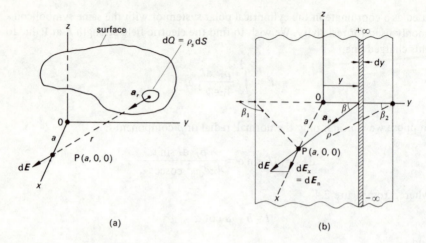

Figure 2.5   (a) *Surface in yz plane.* (b) *Infinitely long plane.*

The surface element can be an infinite strip dy wide having $\rho_L = \rho_S \, dy$. From the previous section

$$dE = \frac{\rho_S \, dy}{2\pi\epsilon_0 \rho} \, a_\rho$$

in which $y = a \cot \beta$; $dy = -a \, \text{cpsec}^2 \, \beta \, d\beta$ and $\rho = a \, \text{cosec} \, \beta$. We are concerned only with the electric field strength E_x in the x direction

$$dE_x = \frac{\rho_S \, dy \sin \beta}{2\pi\epsilon_0 \rho} \, a_x = \frac{\rho_S(-a \, \text{cosec}^2 \, \beta \, d\beta) \sin \beta}{2\pi\epsilon_0 a \, \text{cosec} \, \beta} \, a_x$$

$$E_x = \frac{\rho_S}{2\pi\epsilon_0} \int_{\beta_1}^{\beta_2} (-d\beta)a_x = \frac{\rho_S}{2\pi\epsilon_0} (\beta_1 - \beta_2)a_x$$

For an infinitely wide plate, $\beta_1 \to \pi$, $\beta_2 \to 0$ and finally

$$E_x = \frac{\rho_S}{2\epsilon_0} \, a_x = E_n = \frac{\rho_S}{2\epsilon_0} \, a_n$$

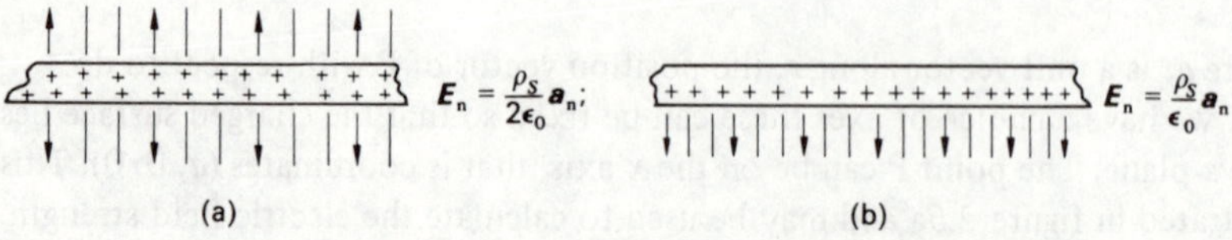

$E_n = \frac{\rho_S}{2\epsilon_0} a_n$; $E_n = \frac{\rho_S}{\epsilon_0} a_n$

(a)   (b)

Figure 2.6   *Field due to uniform surface charge ρ_S coulombs per square metre.*
(a) *Charge on both sides.* (b) *Charge on one side.*

indicating a field normal to the plate. This is surprising for E_n has no dependence upon the distance a from the plate. The electric field strength is everywhere uniformly normal to the plate. In this the charge exists uniformly on *both* faces of the plate as illustrated in figure 2.6a.

Should the charge reside only on one face of the plate (as on one plate of a charged capacitor) the electric field strength is twice as big, see figure 2.6b.

$$E_n \text{ (normal)} = \frac{\rho_S}{\epsilon_0}\, a_n$$

2.2.4 *Field due to a volume charge ρ_V coulombs/m^3*

A volume charge density is shown in figure 2.7 where ρ_V is the charge density at any point A in the volume. If Q is the total charge involved and dQ the charge associated with the differential volume d(vol), then

$$dQ = \rho_V \, d(vol); \qquad Q = \int_{vol} \rho_V \, d(vol); \qquad dE = \frac{\rho_V \, d(vol)}{4\pi\epsilon_0 r^2}\, a_r$$

$$E = \frac{1}{4\pi\epsilon_0} \int_{vol} \frac{\rho_V \, d(vol)}{r^2}\, a_r$$

This is really a triple integral and likely to be difficult; but this difficulty may be reduced, as we shall see later, by the use of scalar potential.

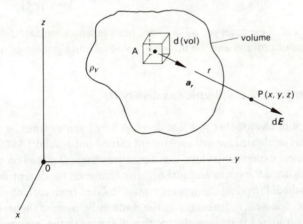

Figure 2.7 *Field due to volume charge ρ_V coulombs per cubic metre.*

A particular charge distribution is indicated in figure 2.8. Here we have a column of charge of cross-section $2a$ $2b$ symmetric about the z axis and extending to $+\infty$ and $-\infty$ along the z direction.

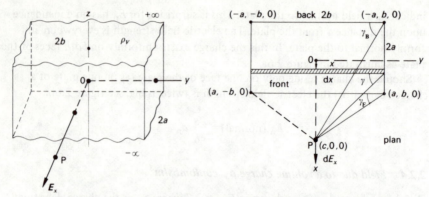

Figure 2.8   *Field outside a column of charge ρ_V coulombs per cubic metre.*

For the rectangular column of charge shown, the normal electric field strength $E_n = E_x$ at a symmetrical point P can be found by considering the elementary wide plate, dx thick, having 'surface' charge density $\rho_V\, dx$. From the previous case we have

$$dE_x = \frac{\rho_V\, dx}{2\pi\epsilon_0}\,(\pi - 2\gamma)a_x$$

and using $x = c - b(\tan \gamma)$ the integral can be evaluated between the limits γ_F and γ_B. The result is

$$E_x = \frac{\rho_V}{\pi\epsilon_0}\left[a(\gamma_F + \gamma_B) + c(\gamma_B - \gamma_F) + b \ln\left(\frac{\cos \gamma_B}{\cos \gamma_F}\right)\right]a_x$$

Of course in all these cases ρ_L, ρ_S, ρ_V have been taken as constants; they could all be functions of position and the integrals would be correspondingly more difficult.

2.3   Electric flux Ψ and electric flux density D

The concept of electric flux was first developed by Faraday when he considered the results of his celebrated ice pail experiment carried out around 1837. Briefly, the results of these experiments show that a positive charge Q placed on the inner sphere of figure 2.9 induces an equal negative charge on the outer sphere which may be earthed. Providing the spheres remain isolated from each other the charge on the outer sphere is independent of the medium between the spheres. It was therefore considered that some *electric flux Ψ* existed in the intervening space between the positive and negative charges. The amount of flux is directly proportional to the charge since a decrease in charge on the inner sphere is accompanied with a corresponding decrease in charge on the outer. In SI units one line or tube of flux emanates from +1 coulomb and terminates on −1 coulomb and $\Psi = Q$ numerically.

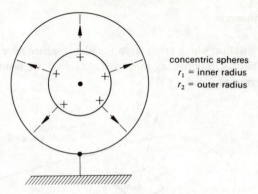

concentric spheres
r_1 = inner radius
r_2 = outer radius

Figure 2.9 *Faraday's 'ice pail' experiment.*

The electric flux density D, which was called displacement by Maxwell, is defined as the electric flux per unit normal area.

If a spherical surface of radius r in figure 2.9, where $r_1 < r < r_2$, is considered, the total electric flux equals Q and the area is $4\pi r^2$

$$D = \frac{\text{total flux}}{\text{area}}\, a_r = \frac{Q}{4\pi r^2}\, a_r = \epsilon_0 E$$

Where the surface is not everywhere normal to the electric displacement the total flux is given by integrating the dot product, D. dS over the whole surface as indicated in figure 2.10.

$$\Psi = \int_S D \, . \, \mathrm{d}S = Q$$

Here it is worth noting the importance of whether the surface is open or closed, as for example in a plate which might have charges on both sides. This leads to the introduction of Gauss's theorem in the next section. For further information on electric flux and Faraday's ice pail experiment, see Baden-Fuller (1973) and Brown (1973).

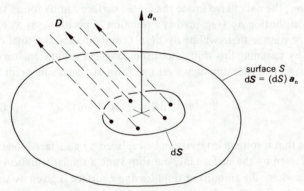

Figure 2.10 *Flux density and surface not orthogonal.*

2.4 Gauss's theorem

This is a generalisation which arises from Faraday's sphere experiment. It states that the total electric flux passing through a *closed surface* is numerically equal to the electric charge enclosed.

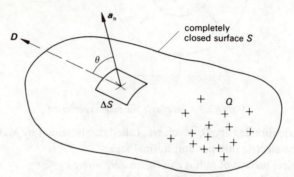

Figure 2.11   *Illustration of Gauss's theorem.*

Consider a completely closed surface S containing charge Q as shown in figure 2.11. If a_n is the unit normal vector to a surface element ΔS then

$$\Delta S = (\Delta S)a_n$$

The electric flux density at the surface element ΔS is D and

$$\Delta \Psi = D \Delta S \cos \theta = D \cdot \Delta S$$

$$\Psi = \oint_S d\Psi = \oint_S D \cdot dS = \epsilon_0 \oint_S E \cdot dS = Q \text{ coulombs}$$

In this Q is the net coulomb charge $(Q_+ - Q_-)$ inside the closed surface. The integral $\oint_S$ means that it must be over a *closed surface S*.

However, the net charge inside the closed surface can be found from the volume density distribution ρ_V (regarded as a function of position, say x, y, z). In each elementary volume there will be ρ_V d(vol) coulombs and the total charge can be obtained by summing this throughout the complete volume enclosed by S. That is $Q = \int_{\text{vol}} \rho_V$ d(vol). Combining we get one form of Gauss's theorem

$$\Psi = \oint_S D \cdot dS = \int_{\text{vol}} \rho_V \text{ d(vol)} = Q \qquad \text{Gauss's theorem}$$

Note here that a volume integral can be replaced by a surface integral. The amount of flux is given by the surface integral above and a similar situation occurs in fluid mechanics where the amount of fluid leaving a surface is given by the surface integral of the fluid flow density.

2.5 Maxwell's first equation

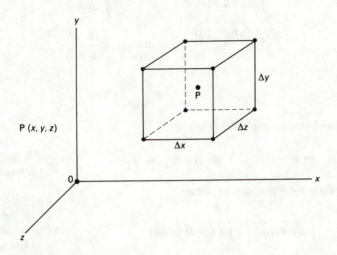

Figure 2.12 *Volume element in an electric field.*

Consider a volume element in an electric field. Gauss's theorem must apply to any closed surface and we proceed to the volume element $\Delta(\text{vol}) = \Delta x \, \Delta y \, \Delta z$ as shown in figure 2.12. At the central point P (x, y, z)

$$D = D_x a_x + D_y a_y + D_z a_z$$

D_x, D_y, D_z are all functions of the position coordinates x, y, z and vary accordingly. The surface integral $\oint_S D \cdot dS$ has to be taken over *six* sides

$$\oint_S D \cdot dS = \int_{\text{front}} + \int_{\text{back}} + \int_{\text{left}} + \int_{\text{right}} + \int_{\text{top}} + \int_{\text{bottom}}$$

$$\int_{\text{front}} = D_{\text{front}} \Delta x \, \Delta y = D_z + \tfrac{1}{2}\Delta z \, \frac{\partial D_z}{\partial z} \, \Delta x \, \Delta y$$

$$\int_{\text{back}} = D_{\text{back}} \Delta x \, \Delta y = D_z - \tfrac{1}{2}\Delta z \, \frac{\partial D_z}{\partial z} \, \Delta x \, \Delta y$$

$$\int_{\text{front}} + \int_{\text{back}} = \frac{\partial D_z}{\partial z} \, \Delta x \, \Delta y \, \Delta z = \frac{\partial D_z}{\partial z} \, \Delta(\text{vol})$$

Repeating with left, right and top, bottom in pairs we can write finally

$$\oint_S D \cdot dS = \left(\frac{\partial D_x}{\partial x} + \frac{\partial D_y}{\partial y} + \frac{\partial D_z}{\partial z} \right) \Delta(\text{vol})$$

$$\frac{\partial D_x}{\partial x} + \frac{\partial D_y}{\partial y} + \frac{\partial D_z}{\partial z} = \lim_{\Delta(\text{vol}) \to 0} \oint_S \frac{D \cdot dS}{\Delta(\text{vol})}$$

$$= \lim \frac{\Delta Q}{\Delta(\text{vol})} = \rho_V$$

The left-hand side is simply div D hence

$$\text{div } D = \nabla \cdot D = \rho_V$$

Maxwell's first equation
(differential form)

If we integrate the equations we obtain

$$\oint_S D \cdot dS = \int_{\text{vol}} \rho_V \, d(\text{vol}) = \int_{\text{vol}} (\nabla \cdot D) \, d(\text{vol})$$

Maxwell's first equation
(integral form)

This is also one form of the *divergence theorem* by means of which the surface integral of D is replaced by the volume integral of div D, or vice versa.

2.6   Energy and potential

We have seen that, to obtain the distribution of the electric field strength E, either Coulomb's law or Gauss's theorem may be used. The first requires a vector summation of the contributions due to each charge; the second is really only applicable to symmetrical charge distributions where a suitable gaussian surface can easily be seen.

Using the scalar potential function avoids the vector summation and only a relatively simple differentiating procedure is required.

2.6.1   Potential difference

Consider the work done in moving a point charge $+q$ in an electric field of strength E which is constant in space as shown in figure 2.13a.

Work done by charge = $qE \cdot dl$. Hence work done *on* charge is $-qE \cdot dl$, that is $dW = -qE \cdot dl$ joules. Now suppose $+q$ is moved from B to A. Then the work done ($W_{B \to A}$) *on* the charge is (see figure 2.13b)

$$W_{B \to A} = -q \int_B^A E \cdot dl \text{ joules} = -qE \int^A dl \cos \theta = -qE(B'A')$$

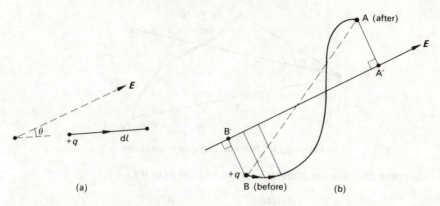

Figure 2.13   *Movement of a test charge +q in an electric field E.*

if E is constant in space. This is clearly independent of the actual path taken between B and A, and depends only on the start and finish.

In general the electric field strength E is not constant in space; even so the work done is still independent of the actual path. In this general case

$$W_{B \to A} = -q \int\limits_{B}^{A} E \,.\, \mathrm{d}l$$

If $q = +1$ coulomb, the work done in moving $+q$ from B to A is known as the potential of A with reference to B, or the potential difference between points A and B

$$V_{AB} = \text{potential of A with reference to B}$$

$$= V_A - V_B = -\int\limits_{B}^{A} E \,.\, \mathrm{d}l \text{ volts}$$

As derived, the potential in volts is the work in joules per coulomb. The electric field strength E is thus also measured in volts per metre which is identical with newtons per coulomb, see section 2.2.

2.6.2   *Potential in field of a single point charge Q*

In this case the electric field is equal to $(Q/4\pi r^2 \epsilon_0)a_r$. In figure 2.14, A and B are two points in such a field and, to find the potential difference between A and B, a test charge $q = +1$ C is moved over any path from B to A and the *work done on it* recorded.

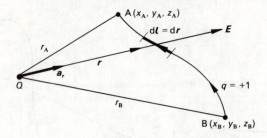

Figure 2.14   *Potential due to a point charge.*

Suppose cartesian coordinates are used and the path B to A is

$$
\begin{array}{cc}
\text{Before} & \text{After}
\end{array}
$$

$$
B\left|\begin{array}{ccc}
x_B \to x_A & x_A & x_A \\
y_B & y_B \to y_A & y_A \\
z_B & z_B & z_B \to z_A
\end{array}\right| A
$$

$$
r = xa_x + ya_y + za_z \qquad dr = dx\,a_x + dy\,a_y + dz\,a_z = dl
$$

$$
V_{AB} = V_A - V_B = -\int_B^A E \cdot dl = -\int_B^A \left(\frac{Q}{4\pi\epsilon_0(x^2 + y^2 + z^2)}\right)\left(\frac{xa_x + ya_y + za_z}{(x^2 + y^2 + z^2)^{1/2}}\right) \cdot dl
$$

$$
= -\frac{Q}{4\pi\epsilon_0}\int_B^A \frac{x\,dx + y\,dy + z\,dz}{(x^2 + y^2 + z^2)^{3/2}}
$$

$$
= -\frac{Q}{4\pi\epsilon_0}\left[\int_{x_B}^{x_A}\frac{x\,dx}{(x^2 + y^2 + z^2)^{3/2}} + \int_{x_B}^{y_A}\frac{y\,dy}{(x^2 + y^2 + z^2)^{3/2}} + \int_{z_B}^{z_A}\frac{z\,dz}{(x^2 + y^2 + z^2)^{3/2}}\right]
$$

$$
= +\frac{Q}{4\pi\epsilon_0}\{|(x^2 + y^2 + z^2)^{-1/2}|_{x_B}^{x_A} + |(x^2 + y^2 + z^2)^{-1/2}|_{y_B}^{y_A}
$$
$$
+ |(x^2 + y^2 + z^2)^{-1/2}|_{z_B}^{z_A}\}
$$

$$
= \frac{Q}{4\pi\epsilon_0}\left[(x_A^2 + y_B^2 + z_B^2)^{-1/2} + (x_A^2 + y_A^2 + z_B^2)^{-1/2}\right.
$$
$$
+ (x_A^2 + y_A^2 + z_A^2)^{-1/2} - (x_B^2 + y_B^2 + z_B^2)^{-1/2} - (x_A^2 + y_B^2 + z_B^2)^{-1/2}
$$
$$
\left. - (x_A^2 + y_A^2 + z_B^2)^{-1/2}\right]
$$

$$
= \frac{Q}{4\pi\epsilon_0}\{(x_A^2 + y_A^2 + z_A^2)^{-1/2} - (x_B^2 + y_B^2 + z_B^2)^{-1/2}\}
$$

$$
= \frac{Q}{4\pi\epsilon_0}\left(\frac{1}{r_A} - \frac{1}{r_B}\right) \text{ volts}
$$

Of course this expression could have been produced far more easily by using spherical coordinates, for the field of a single point charge Q has spherical symmetry. The choice of the right coordinates plays an important part in problem solving.

If the reference point B is at infinity we can take $r_B \to \infty$ and assume $V_B = 0$.

$$V_A = \text{potential of A with reference to } V_B = 0 \text{ at infinity}$$

$$= \frac{Q}{4\pi\epsilon_0} \frac{1}{r_A} \text{ volts}$$

Similarly for any other point C the potential V_C is $(Q/4\pi\epsilon_0)(1/r_C)$ and $V_{AC} = V_A - V_C = (Q/4\pi\epsilon_0)(1/r_A - 1/r_C)$.

2.6.3 *Potential due to other charge distributions*

It is now possible to consider the potential produced at a point by various types of charge distributions, and examples are given of four simple distributions.

(i) A system of point charges Q_1 at r_1, Q_2 at r_2, . . ., Q_n at r_n

$$\frac{1}{4\pi\epsilon_0} \sum_{n=1}^{n} \frac{Q_n}{r_n}$$

(ii) A line charge ρ_L per unit length

$$\frac{1}{4\pi\epsilon_0} \int_L \frac{\rho_L}{r} \, dr$$

(iii) A surface charge ρ_S per unit area

$$\frac{1}{4\pi\epsilon_0} \int_S \frac{\rho_S}{r} \, dS$$

(iv) A volume charge ρ_V per unit volume

$$\frac{1}{4\pi\epsilon_0} \int_{vol} \frac{\rho_V \, d(\text{vol})}{r}$$

In each an assemblage of charges is considered and a summation of the potential contributions is made simple, for potential is a scalar quantity. The distance of the point from each particular element considered is r. In general these integrals are far more easily evaluated than the corresponding ones for electric field strength E.

2.7   Conservative field

In the scalar potential field V, no work can be involved in carrying a test charge ($+1$ C) round any closed path, such as that shown in figure 2.15. Since the potential difference $V_A - V_B$ is independent of the path $V_{AB} = -V_{BA}$ and clearly

$$V_{AB} + V_{BA} = 0$$

But this is also the integral of $-E.\,dl$ taken round the path so that finally

$$V_{AB} + V_{BA} = \oint E.\,dl = 0$$

Fields having this property are said to be *conservative*; this applies to the electric field strength E as well as V. We may say generally that, whenever a vector field F is derived from a scalar potential field S, then the field F is conservative and $\oint F.\,dl = 0$. Conservative fields have already been mentioned in section 1.5.

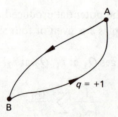

Figure 2.15   *A closed path.*

Of course, if work were involved in moving a charge round the closed path $B \rightarrow A \rightarrow B$, then it should be possible to contrive a continuous supply of energy!

2.7.1   *Maxwell's second equation*

In the conservative electric field of strength E, derived from the scalar potential V, we have round any closed path

$$\oint_l E.\,dl = 0$$

Maxwell's second equation
(integral form)

The meaning and definition of *curl* of a vector has already been discussed

$$\operatorname{curl} F = \lim_{\Delta S \to 0} \frac{\oint F.\,dl}{\Delta S}\, a_n = \nabla \times F$$

and also ΔS is oriented to give the greatest value of $\oint F.\,dl$. Since $\oint_l E.\,dl = 0$, the electrostatic field can have no curl.

$$\operatorname{curl} E = \nabla \times E = 0$$

Maxwell's second equation
(differential form)

2.8 Potential gradient

If a test charge ($q = +1$ C) is moved a distance Δl in an electric field of strength E, which has a scalar potential V, then $\Delta V = -E \cdot \Delta l$. This change will be biggest when the step Δl is made in the same direction as E; that is when $\cos \theta = 1$ along a_n; see figure 2.16.

$$E = - \left| \frac{\Delta V}{\Delta l} \right|_{\max} a_n \rightarrow - \left| \frac{dV}{dl} \right|_{\max} a_n$$

If Δl is made along one of the equipotentials, then $\Delta V = -E \cdot \Delta l = 0$ for $\theta = 90°$. Clearly E is directed along a_n, the normal to the equipotential surfaces V. The

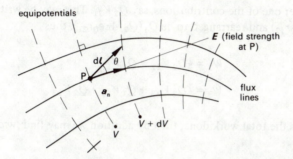

Figure 2.16 *A two-dimensional plot of potential gradient in an electric field E.*

operation on V by which the maximum value is obtained has already been discussed in section 1.5.1 as the gradient of a scalar

$$\text{grad } V = \nabla V = \frac{\partial V}{\partial x} a_x + \frac{\partial V}{\partial y} a_y + \frac{\partial V}{\partial z} a_z = -E$$

2.9 The potential energy of a system of point charges

Here it is necessary to find the work done in establishing the set of point charges in a region with the zero of potential at infinity; this work is equal to the stored energy. Consider, for example, three point charges Q_1, Q_2, Q_3 to be set up at positions 1, 2, 3, as shown in figure 2.17. The notation for potential is that V_{32}

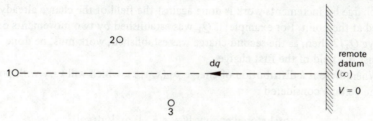

Figure 2.17 *A set of three point charges.*

refers to the potential at point 3 due to charge Q_2 at point 2. Assuming that no other charges are present and that the charges will be established in the order Q_1, Q_2, Q_3, then the work done W_1 in establishing Q, in small increments, is $\frac{1}{2}Q_1 V_{11}$ since the average potential at position 1 in establishing Q_1 incrementally is $\frac{1}{2}V_{11}$.

The work done W_2 in establishing Q_2, also in small increments, is

$$W_2 = \tfrac{1}{2}Q_2 V_{22} + Q_2 V_{21}$$

and for Q_3 is

$$W_3 = \tfrac{1}{2}Q_3 V_{33} + Q_3 V_{31} + Q_3 V_{32}$$

Now consider one of the contributions, say $Q_2 V_{21}$. This may be written $Q_2(Q_1/4\pi\epsilon_0 r_{12})$ and rearranged to be $Q_1(Q_2/4\pi\epsilon_0 r_{21})$, thus

$$W_2 = \tfrac{1}{2}Q_2 V_{22} + Q_1 V_{12}$$

$$W_3 = \tfrac{1}{2}Q_3 V_{33} + Q_1 V_{13} + Q_2 V_{23}$$

and, if W_E is the total work done, then by addition we may find *twice* the work done

$$2W_E = Q_1 V_{11} + Q_2 V_{22} + Q_3 V_{33} + Q_1 V_{12} + Q_2 V_{21} + Q_1 V_{13}$$
$$+ Q_2 V_{23} + Q_3 V_{31} + Q_3 V_{32}$$
$$= Q_1(V_{11} + V_{12} + V_{13}) + Q_2(V_{21} + V_{22} + V_{23}) + Q_3(V_{31} + V_{32} + V_{33})$$
$$= Q_1 V_1 + Q_2 V_2 + Q_3 V_3$$

where V_1, V_2, V_3 are the potentials at positions 1, 2 and 3. Extending this to any number n of charges

$$W_E = \tfrac{1}{2}(Q_1 V_1 + Q_2 V_2 + \ldots Q_n V_n)\ \text{joules}$$

In the above analysis the terms such as $\frac{1}{2}Q_1 V_{11}$ arise because the charge has been established in increments and, in the absence of other charges, a discrete charge Q_1 could be established in one operation without doing work. However, in establishing the charge by increments work is done against the field of the charge already established at the point. For example, if Q_1 was established by two movements of charge $Q_1/2$ then, as the second charge was established, work must be done against the field of the first charge.

When there is a distribution of charge $\rho_V(x, y, z)$, which will vary with position in the volume considered

$$\text{total stored energy}\ W_E = \tfrac{1}{2} \int_{\text{vol}} \rho_V V\, \mathrm{d(vol)}$$

We can now re-interpret this stored energy in terms of the electric displacement and electric field strength rather than in terms of charge and potential.

2.10   Energy density in an electric field

Starting from the above expression for the stored energy of a charge distribution

$$W_E = \tfrac{1}{2} \int_{vol} \rho_V V \, d(vol)$$

for the charge system, we replace $\rho_V = \text{div } D = \nabla . D$ and get

$$W_E = \tfrac{1}{2} \int_{vol} (\nabla . D) V \, d(vol)$$

Now consider the vector identity relation

$$\nabla . (SF) = S \nabla . F + \nabla S . F$$

and rearrange it as

$$(\nabla . F)S = \nabla . (FS) - F . \nabla S$$

Replacing V for S and D for F the energy relation becomes

$$W_E = \tfrac{1}{2} \int_{vol} (\nabla . DV) \, d(vol) - \tfrac{1}{2} \int_{vol} (D . \nabla V) \, d(vol)$$

The divergence theorem enables us to replace a volume integral by a surface integral as follows

$$\left(\int_{vol} \rho_V \, d(vol) \right) = \int_{vol} (\nabla . D) \, d(vol) = \int_S D . dS \qquad \begin{array}{l} \text{divergence} \\ \text{theorem} \end{array}$$

This can be used on the first of the two volume integrals giving

$$W_E = \tfrac{1}{2} \int_S (DV) . dS - \tfrac{1}{2} \int_{vol} (D . \nabla V) \, d(vol)$$

The first of these integrals is zero for, taking the surface approaching infinity, D is proportional to $1/r^2$, V is proportional to $1/r$ and S is proportional to r^2; the integral therefore tends to zero.

Finally we are left with the second volume integral in which we can replace ∇V by $-E$ giving

$$W_E = \tfrac{1}{2} \int_{vol} D . E \, d(vol) = \tfrac{1}{2} \int_{vol} \epsilon_0 E^2 \, d(vol)$$

On a differential basis, $dW_E/d(\text{vol}) = $ energy density $= \frac{1}{2}D \cdot E$ joules/metre^3. It is a simple step to associate this energy dW_E with the volume $d(\text{vol})$, though it strictly cannot be said that the energy does reside in the fields in the volume any more than it may reside in the charges and potentials at the surfaces which produce the fields.

Problems

2.1 In a 1 m cube lying between $(0, 0, 0)$ and $(1, 1, 1)$ the charge density is given by $-(x^2 + y^2 + z^2)^{2 \cdot 5}$ C/m^3. Show that the electric field strength E at the origin is $(7/48\pi\epsilon_0)(a_x + a_y + a_z)$ V/m.

2.2 An infinite uniform line charge $\rho_L = 2$ nC/m lies on a line defined by $r \times a_y = 3a_z$, where $r = xa_x + ya_y + za_z$ from the origin to a point on the line. Find the electric field strength E at $(0, 5, 0)$ and at $(1, 2, 6)$.

$(-12a_x, 9/5(-a_x + 3a_z)$ V/m$)$

2.3 An infinite sheet with surface charge $\rho_S = 12\epsilon_0$ C/m^2 is lying in the plane $x - 2y + 3z = 4$. Find an expression for the electric field strength E on the side of the plane including the origin.

$(-6(a_x - 2a_y + 3a_z)/\sqrt{14}$ V/m$)$

2.4 (a) A line charge of 1 nC/m in a ring of 0·5 m radius is centred at $(0, 0, 0)$ in the $z = 0$ plane. Use a gaussian surface to estimate the electric field strength E at the points $(0·51, 0, 0)$ and $(10, 10, 10)$.
(b) Each metre of a long straight source emits electrons at $6·24 \times 10^{18}$ per second at a constant radial velocity of 10^7 m/s. Find expressions for the electric field strength E and volume charge density ρ_V at a general point (ρ, ϕ, z). Comment on the assumption of constant electron velocity, for a cylindrical space charge of 10 cm radius.

$((a)$ $1·8a_r$ kV/m, $0·0545(a_x + a_y + a_z)$ V/m;

(b) $-1800a_\rho$ V/m, $-1·593/\rho \times 10^8$ C/m^3)

2.5 (a) Given an electric field strength $E = (y + 1)a_x + (x - 1)a_y + 2a_z$ volts/m, find the potential differences V_{AB} and V_{CD} if A, B, C, D are $(2, -2, -1)$, $(0, 0, 0)$, $(3, 2, -1)$ and $(-2, -3, +4)$.
(b) Given $F = xya_x + x^2a_y + 0a_z$, evaluate $\oint F \cdot dl$ round the two closed paths $(1, 1, 0)(-1, 1, 0)(-1, -1, 0)(1, -1, 0)(1, 1, 0)$ and $(0, 0, 0)(1, 0, 0)$ $(1, 1, 0)(0, 1, 0)(0, 0, 0)$. Is F a conservative field? Check by curl F.

$((a)$ + 2 V, +10 V; (b) 0, $+\frac{1}{2}$, no$)$

2.6   Find the electric field strength E and volume charge density ρ_V at the point (a, a, a) for each of the fields in which scalar V is given as follows.

(a) $V_0(x + y)(x + z)/a^2$,
(b) $(V_0/a)\rho \cos \phi \exp(-z/a)$,
(c) $V_0 a^2 \sin \theta \cos \phi/r^2$.

((a) $(-2V_0/a)(2a_x + a_y + a_z), -2\epsilon_0 V_0/a^2$,
(b) $V_0[-a_r + a_\phi + (2)^{1/2}a_z]/(2)^{1/2}a \exp(1), -\epsilon_0 V_0/a^3 \exp(1)$,
(c) $(V_0/18a) [4a_r - (2)^{1/2}a_\theta + (6)^{1/2}a_\phi], 0)$

2.7   Find the energy stored in the region $r > a$ for a point charge Q at the origin.

$((1/2\epsilon_0)(Q^2/4\pi a))$

2.8   Given an electric field strength $E = (p/4\pi\epsilon_0 r^3)(2 \cos \theta a_r + \sin \theta a_\theta)$ V/m for a dipole p in the z direction at the origin, find the energy stored in the region $r > a$. Comment on the result when $a \to 0$.

$((1/288)(p^2/\epsilon_0 a^3)(16 + 3\pi))$

2.9   (a) An infinitely long straight conductor is charged $-\rho_L$ C/m^2 and passes through $(-a, 0)$ parallel to the z axis as shown in figure P2.9a. The potential V at any point $P(x, y)$ may be taken as $V_0 \ln\{(x + a)^2 + y^2\}$ volts. Find expressions for the electric field strength E, and volume charge density ρ_V at point P.

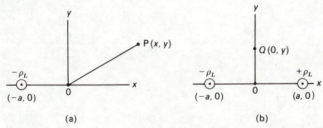

Figure P2.9.

(b) As shown in figure P2.9b, a similar parallel line charged ($+ \rho_L$) has been added through the point $(+a, 0)$. Deduce an expression for the electric field strength E_Q at any point $Q(0, y)$ on the y axis. If E_Q must not exceed 500 kV/m, find the greatest permissible V_0 for $a = 1$.

((a) $-V_0[(x + a)^2 + y^2]^{-1} [2(x + a)a_x + 2ya_y], 0$
(b) $-4V_0 a(a^2 + y^2)^{-1}a_x, < 125$ kV)

2.10   Three charges of values $4\pi\epsilon_0(+4, -3, +5)$ coulombs are established on conducting spheres each of 10 cm diameter centred at points $(30, 0, 0)$ $(0, 40, 0)$ $(0, 0, 50)$ cm.

(a) Find the individual potentials of the three spheres (with reference to $V = 0$ at infinity) and the total stored energy.

(b) What is the electric field strength E at the origin?

Assume the charges are uniformly distributed on the sphere surfaces.

((a) $82 \cdot 71, -42 \cdot 62, 102 \cdot 42, 2\pi\epsilon_0(970 \cdot 8)$ joules; (b) $-44 \cdot 4a_x + 18 \cdot 75a_y - 20a_z$)

3

The Electric Field and Materials

In the previous chapter we were concerned with the electrostatic field and effect of charge distributions in free space. We now consider electric fields in the presence of materials. The space in which an electric field exists is usually a dielectric such as air, mica, oil, etc., and the boundaries of this space where charges in the main reside are conductors such as aluminium, copper, etc.

Conduction in materials is by the drift of electrons under the influence of an electric field and/or by means of diffusion in the presence of electron concentration gradients. This latter process occurs of course in the absence of an electric field. Each electron carries a charge $-q_e$ of -1.6×10^{-19} coulombs. There are three types of material

(i) Conductors in which there are many electrons available for electrical conduction and in which the resistance to current flow increases as the temperature increases.
(ii) Semiconductors in which the conduction process is by means of electrons and holes (electron vacancies with a charge of $+1.6 \times 10^{-19}$ coulombs), and in which an increase in temperature causes a decrease in resistance.
(iii) Insulators (dielectrics) in which there are no electrons available for the conduction process until large electric fields cause breakdown in the electronic structure of the atoms of the material.

It is not intended to cover the topic of electronic conduction in solids in any great detail in this text and for further information the reader is referred to books on electrical materials (Solymar and Walsh, 1970; Allison, 1971).

3.1 Current

Charges in motion constitute an electric current; the material in which they move

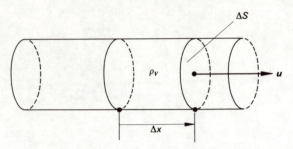

Figure 3.1　*Flow of charge in a conducting filament.*

is a conductor. Current I can be defined

$$I = \frac{\text{charge}}{\text{time}} = \frac{Q}{t} = \frac{\mathrm{d}Q}{\mathrm{d}t}$$

Suppose a conductor filament of area ΔS, as shown in figure 3.1, has a flow of (positive) charge of density ρ_V (C/m^3) at velocity u (m/s). In time Δt

charge passing surface　$\Delta Q = \rho_V \, \mathrm{d(vol)} = \rho_V \Delta S \Delta x$

current in filament　$\Delta I = \dfrac{\Delta Q}{\Delta t} = \dfrac{\rho_V \Delta S \Delta x}{\Delta t} = \rho_V u \Delta S$

Hence

current density $J = \dfrac{\Delta I}{\Delta S} = \rho_V u$

In the above the direction of u and J is clearly fixed by the filament. In an extensive body where the charge flow may be more random, J and u are similarly oriented vectors

$$J = \rho_V u$$

and through a surface S the current (a scalar) will be

$$I = \int_S J \cdot \mathrm{d}S = \int_S \rho_V u \cdot \mathrm{d}S$$

The dot product needed for $\mathrm{d}S$ may *not* be orthogonal to the charge velocity u.

3.2　The continuity equation

The total current which is emerging from a closed surface is

$$I_{\text{out}} = \oint_S J \cdot \mathrm{d}S = -\frac{\mathrm{d}}{\mathrm{d}t} Q_{\text{inside}}$$

where Q_{inside} is the total charge inside the closed surface. In circuit analysis it is usual (with capacitors) to take the current as flowing *into* the surface of the capacitors so that there

$$I_{into} = + \frac{\mathrm{d}}{\mathrm{d}t} Q_{inside}$$

We shall here continue to take *outward* current flow in the above equation; we can employ the divergence theorem to change the surface integral to a volume integral

$$I_{out} = \oint_S \boldsymbol{J} . \, \mathrm{d}\boldsymbol{S} = \int_{vol} (\nabla . \boldsymbol{J}) \, \mathrm{d}(vol)$$

$$= -\frac{\mathrm{d}}{\mathrm{d}t} Q_{inside} = -\frac{\mathrm{d}}{\mathrm{d}t} \int_{vol} \rho_V \mathrm{d}(vol)$$

From these

$$\nabla . \boldsymbol{J} = \mathrm{div}\, \boldsymbol{J} = -\frac{\partial \rho_V}{\partial t} \qquad \qquad \text{continuity equation}$$

This outward current flow is indicated in figure 3.2.

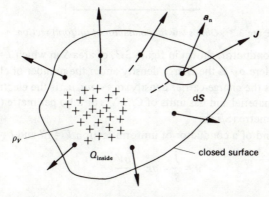

Figure 3.2   *Current from an extensive, charged body.*

The current or charge per second diverging from a unit volume is equal to the time rate of decrease of charge per unit volume. This equation states that there should be no accumulation of charge and is of great importance in semiconductor-device theory where it is used to describe the variation in time of the charge-carrier density within a particular volume element.

3.3   Metallic conductors

In a metal there are many valence electrons which may be moved through the material under the action of an electric field. These 'free' electrons each have a charge of $-q_e$, and the mechanical force on each electron will be $-q_e E$, where E is the applied field strength. However, unlike a truly free electron the electrons inside the metal have a complex motion within the crystalline lattice structure of the material. One of the consequences of electron motion in a periodic lattice potential is the appearance of an 'effective mass' m^* which is vastly different from the rest mass of an electron. Another result is that under the action of an applied field E the electron moves with a velocity which is proportional to E. This velocity u_d is the drift velocity and, for electrons

$$u_d = -\mu_n E$$

where μ_n is the mobility and is measured in m^2/V s. For holes the charge is $+q_e$ and the mobility μ_p, in general for a material in which electrons and holes are present $\mu_n \neq \mu_p$.

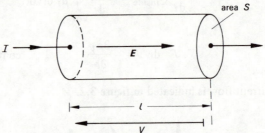

Figure 3.3 *Ohm's law for conductor of uniform section.*

Consider the conductor, shown in figure 3.3, of area S in which $J = \rho_V u = -nq_e \mu_n E = \sigma E$. Here ρ_V is the charge density and n the number of electrons per m^3 which is known as the charge-carrier density. $\sigma = -nq_e \mu_n$ is the electrical conductivity of the material and has units of C/m V s or mho per metre $(\Omega \, m)^{-1}$, or now siemens per metre (S/m).

From end to end of a conductor of uniform section $V = El$ and $I = JS$ so that

$$J = \frac{I}{S} = \sigma E = \frac{\sigma V}{l}$$

$$V = I \frac{l}{\sigma S} = IR$$

The two equations $J = \sigma E$ and $V = IR$ are two forms of Ohm's law; $1/\sigma$ is often spoken of as the *resistivity* of the conductor material, and is written as ρ.

Typical values of mobility and conductivity are given in table 3.1.

Table 3.1 *Mobility and conductivity in some materials*

Material	Mobility (m^2/Vs)	Conductivity (S/m)
Al	0·0014	$3·5 \times 10^7$
Cu	0·0032	$5·8 \times 10^7$
Ag	0·0052	$6·1 \times 10^7$
Ge (intrinsic)	μ_n 0·38, μ_p 0·19	2·2
Si (intrinsic)	μ_n 0·12, μ_p 0·048	$1·6 \times 10^{-3}$

3.4 Boundary conditions in ideal conductors

(i) Any charges of like sign liberated within a conductor appear at the surface.

(ii) No charge may remain *within* the body of a conductor, and neither can an electric field exist inside.

(iii) Charges appearing at the surfaces are influenced by the fields external to the conductor and near it.

(iv) There can be no tangential electric field components at a conductor surface, for this would lead to a charge redistribution until the tangential field was nullified, or the charges vanished.

(v) As there is no tangential component of electric field strength E (and flux density D) the electric flux must leave the conductor surfaces normally so that at the surface

$$D = D_n = \rho_S \cdot a_n$$

To establish these relations consider figure 3.4.

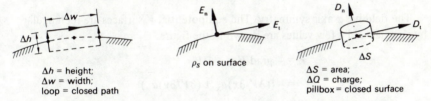

Δh = height;
Δw = width;
loop = closed path

ρ_S on surface

ΔS = area;
ΔQ = charge;
pillbox = closed surface

Figure 3.4   *To establish conditions at conducting boundaries.*

$$\oint E \cdot dl = 0$$

$$= E_n \tfrac{1}{2}\Delta h + E_t \Delta w - E_n \tfrac{1}{2}\Delta h - 0\Delta w$$

$$\oint_S D \cdot dS = Q$$

$$D_n \Delta S = \Delta Q$$

Hence

$$E_t = 0; \qquad D_t = 0$$

Hence

$$D_n = \frac{\Delta Q}{\Delta S} = \rho_S$$

Summarising for a perfect conductor, the following is obtained.

(a) The electric field strength E inside the conductor = 0.

(b) The electric field strength E is external and normal to the surface.

(c) The surface of the conductor *must* be an equipotential.

3.4.1 A boundary-problem example

A two-dimensional potential field $V = 100xy$ is shown in figure 3.5, where surface S is a conductor and which passes through A $(1, -1, 0)$. Just off the surface the potential rises. Sketch the field and find the surface density at A.

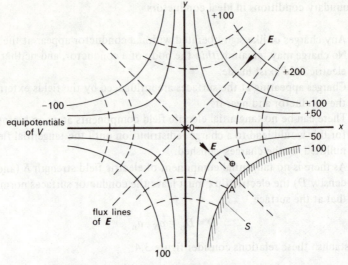

Figure 3.5 *Two-dimensional field sketch*; $V = 100xy$.

The field has z axis symmetry. The equipotential V surfaces are rectangular hyperbolae and a few values are shown in the figure.

$$E = -\text{ grad } V = -\nabla V$$
$$= -((\partial V/\partial x)a_x + (\partial V/\partial y)a_y)$$
$$= -100(ya_x + xa_y) \text{ V/m}$$

The flow lines of the electric field can be sketched in orthogonally to the potential V surfaces. Since the potential off S rises, the conductor lies below A, as shown. The surface of the conductor is an equipotential. The electric field E must be normal to the surface S through A. Hence

$$D = \epsilon_0 E = -100\epsilon_0(ya_x + xa_y)$$
$$= -100\epsilon_0(x^2 + y^2)^{1/2}a_n$$

The normal flux density at A is

$$D_{nA} = \rho_{SA}$$
$$= -\frac{100(2)^{1/2}}{36\pi \times 10^9} = -1 \cdot 25 \text{ nC/m}^2$$

3.5 Dielectrics

Previously in this chapter we have considered conductors in which the charges move freely in response to an electric field so that there is no electric field within the material. Now we deal with insulators (dielectrics) which mostly are concerned with bound charges instead of charges free to move under the action of an electric field. First consider the electric dipole.

3.5.1 The electric dipole

The bound charges in the atoms of dielectric materials can be considered in pairs (+ and −) which have been slightly displaced from the normal charge centres by the action of an electric field. This is the concept of electric dipoles.

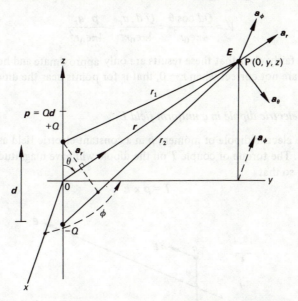

Figure 3.6 *Field E due to a distant dipole.*

Suppose the dipole lies along the z axis, $\pm Q$ charges having separated by d, as shown in figure 3.6. Consider the electric field strength at P $(0, y, z)$. It will be the same for all ϕ. The potential V will be that due to the difference between potentials arising from $+Q$ and $-Q$.

$$V = \frac{Q}{4\pi\epsilon_0}\left(\frac{1}{r_1} - \frac{1}{r_2}\right) = \frac{Q}{4\pi\epsilon_0}\frac{r_2 - r_1}{r_1 r_2} = \frac{Q\,d\cos\theta}{4\pi\epsilon_0 r^2}$$

$$E = -\operatorname{grad} V = -\nabla V = -\frac{Qd}{4\pi\epsilon_0}\nabla(\cos\theta/r^2) \qquad \text{spherical}$$

In spherical coordinates

$$\nabla V = \frac{\partial V}{\partial r}a_r + \frac{1}{r}\frac{\partial V}{\partial \theta}a_\theta + \frac{1}{r\sin\theta}\frac{\partial V}{\partial \phi}a_\phi$$

$$\nabla(\cos\theta/r^2) = -2\frac{\cos\theta}{r^3}a_r - \frac{1}{r}\frac{\sin\theta}{r^2}a_\theta + 0a_\phi$$

Hence

$$E = \frac{Qd}{4\pi\epsilon_0 r^3} \, (2\cos\theta a_r + \sin\theta a_\theta + 0 a_\phi)$$

$$= \frac{p}{4\pi\epsilon_0 r^3} \, (2\cos\theta a_r + \sin\theta a_\theta) = E_r a_r + E_\theta a_\theta$$

In this we have introduced the dipole moment $p = Qd$. It has the direction of d and the separation $-Q$ to $+Q$. It can also be used in the expression for potential V.

$$V = \frac{Qd\cos\theta}{4\pi\epsilon_0 r^2} = \frac{Q\,d.a_r}{4\pi\epsilon_0 r^2} = \frac{p.a_r}{4\pi\epsilon_0 r^2}$$

It should be fairly clear that these results are only approximate and hold when $r \gg d$. They are not correct when $r \to 0$, that is for points near the dipole itself.

3.5.2   The electric dipole in a uniform field E

Consider the electric dipole of moment p in a constant electric field as illustrated in figure 3.7. The torque or couple T on the dipole will have magnitude $qE\,d\sin\theta$; it is a vector so that

$$T = p \times E$$

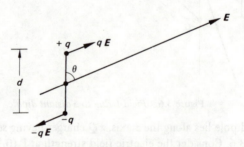

Figure 3.7   *Electric dipole in uniform field.*

As drawn T is directed into the plane of the paper; T tends to decrease the angle θ. Suppose W_p = potential energy of the dipole in field, then $dW_p/d\theta$ is the couple tending to increase θ.

$$W_p = \int T \, d\theta = \int pE\sin\theta \, d\theta = -pE\cos\theta + \text{constant}$$

Choose the constant so that $W_p = 0$ when $\theta = \pi/2$ and finally

$$W_p = -p.E$$

3.5.3   Dielectric polarisation

This can be produced by the action of an electric field E in two ways.

(i) Polar molecules such as HCl, H_2O have permanent dipoles which are randomly oriented; under the action of an electric field they re-align themselves so that there is a resultant dipole moment $p = Qd$ in the direction of the electric field strength E. This orientation is indicated in figure 3.8.

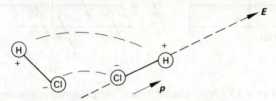

Figure 3.8   *Polar molecule in an electric field.*

(ii) Non-polar molecules do not possess dipoles until the electric field is applied. Then the $+$, $-$ charge centres in the atom move slightly, creating dipoles which align themselves with the direction of the electric field strength E.

Both types of dipole are able to store energy as we have already seen. Both also give rise to polarisation P, on the application of an electric field.

If there are n dipoles in a volume 1 m^3 of dielectric then P = total dipole moment per unit volume = $np = nQd$. Thus for a general volume (vol) in which there are N dipoles and $n = N/(\text{vol})$

$$\text{polarisation } P = (Q_1 d_1 + Q_2 d_2 + \ldots + Q_N d_N)/(\text{vol})$$

$$= (p_1 + p_2 + \ldots + p_m + \ldots + p_N)/(\text{vol})$$

This is a *vector* summation. By and large, the polarisation has been produced by the action of the electric field and we can assume proportionality and write

$$P = \sum_{m=1}^{N} p_m /(\text{vol}) = \chi_e \epsilon_0 E$$

In this χ_e is the proportionality constant, called electric susceptibility (chi_e). There is an analogous constant, χ_m for the magnetic field.

Consider a surface ΔS wholly within a dielectric such as is indicated in figure 3.9a in zero field and in an applied field of strength E in figure 3.9b. The field strength is at an angle θ to the normal to the surface. If the displacement of the positive and negative charges in each molecule is d, then upon applying the field all molecules within $(d/2) \cos \theta$ of the surface will contribute to charge movement across ΔS. Thus molecules 1 and 2 contribute to positive charge crossing the surface and molecules 3 and 4 contribute negative charge. Molecules outside $\pm (d/2) \cos \theta$, such as 5 and 6, do not contribute to charge movement through the surface.

$$\text{Net positive charge } out \text{ of surface} = nQd \cdot \Delta S = \Delta Q_b$$

where subscript b refers to bound charge. Hence

$$\text{net positive charge going } into \text{ the surface} = Q_b = - \oint_S P \cdot dS = \int_{\text{vol}} \rho_b \, d(\text{vol})$$

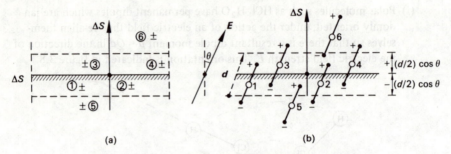

Figure 3.9 *Movement of 'bound' charges during polarisation.*

This is the increase in positive charge within the total volume. We can now apply the divergence theorem

$$Q = \Psi = \int_{\text{vol}} \rho_V \, \text{d(vol)} = \oint_S D . \, \text{d}S = \int_{\text{vol}} (\nabla . D) \, \text{d(vol)}$$

and in terms of bound charge and polarisation

$$Q_b = \int_{\text{vol}} \rho_b \, \text{d(vol)} = - \oint_S P . \, \text{d}S = - \int_{\text{vol}} (\nabla . P) \, \text{d(vol)}$$

and

$$\text{div } P = \nabla . P = -\rho_b$$

This means that the bound charge density ρ_b is given by the *negative* of div P. Apart from the sign change the relation is similar to Maxwell's equation div $D = \rho_V$ and we expect D and P to be closely related derived vectors.

In the presence of a dielectric there are generally both bound charges ρ_b and free charges ρ_f present. These should be related to the basic electric field strength vector E following Coulomb's law and hence

$$\nabla . \epsilon_0 E = \rho_f + \rho_b = \rho_f - \nabla . P$$

and

$$\nabla . (\epsilon_0 E + P) = \rho_f$$

Taking $\epsilon_0 E + P = D$ as the more general form of the displacement vector we now have *two* divergence relations

$$\text{div } \epsilon_0 E = \nabla . \epsilon_0 E = \rho_f + \rho_b \quad \text{(free and bound)}$$

$$\text{div } D = \nabla . (\epsilon_0 E + P) = \rho_f \quad \text{(free only)}$$

There is already a further relation linking P with E, that is

$$P = \chi_e \epsilon_0 E$$

where χ_e is the electric susceptibility; therefore

$$D = \epsilon_0 E + P = \epsilon_0 E(1 + \chi_e) = \epsilon_r \epsilon_0 E = \epsilon E$$

and the expression for the susceptibility becomes

$$\chi_e = \epsilon_r - 1 \qquad \text{and} \qquad \epsilon = \epsilon_r \epsilon_0 = (\chi_e + 1)\epsilon_0$$

The symbol ϵ_r is for the relative permittivity or dielectric constant of the material. Typical values are given in table 3.2. The values quoted in table 3.2 are for the low-frequency, d.c. or electrostatic case only. Many materials will have a marked variation of dielectric constant with frequency of applied field and temperature.

Table 3.2

Material	Dielectric constant
O_2 (1 atmosphere, 25 °C)	1·0005
mineral oil	~2·5
quartz	3·8
mica	5·4
pure H_2O	81
TiO_2	100

Materials for which the simple relationship $P = \epsilon_0 \chi_e E$ holds are called linear isotropic dielectrics. However, many crystalline materials will have different dielectric properties according to different crystal directions and in general the polarisation P is not parallel to the electric field strength E. The components of P in rectangular coordinates become

$$P_i = \sum_j \chi_{e,ij} E_j$$

where i and j refer to the directions x, y, z. Hence the susceptibility has nine components which are referred to as a *tensor*. In this text, however, all dielectrics will be assumed to be linear isotropic materials.

3.5.4 Boundary conditions for dielectrics

We have previously considered the conditions at the boundary between a conductor and an insulator. Now we wish to consider the conditions at the boundary between two different dielectrics. To establish relations, as before we consider a loop integral of the electric field strength E and a gaussian 'pillbox' surface of electric flux density D at the boundary between the two dielectrics illustrated in figure 3.10. The suffixes n and t refer to normal and tangential components.

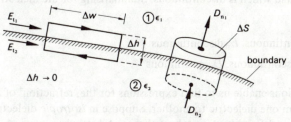

Figure 3.10 *Boundary conditions between two dielectrics.*

Taking the loop integral of the electric field strength E as shown, and $\Delta h \to 0$

$$\oint E \cdot dl = E_{t_1} \Delta w - E_{t_2} \Delta w = 0$$

Hence

$$E_{t_1} = E_{t_2}$$

That is the tangential electric field strength component E_t undergoes no change in passing through the interface and it is said to be *continuous* at the boundary.

Because $D_{t_1} = \epsilon_1 E_{t_1}$ and $D_{t_2} = \epsilon_2 E_{t_2}$ it follows that D_t is discontinuous at the interface where $\epsilon_1 \neq \epsilon_2$.

Thus

$$E_{t_1} = E_{t_2} = \frac{D_{t_1}}{\epsilon_1} = \frac{D_{t_2}}{\epsilon_2}$$

and

$$\frac{D_{t_1}}{D_{t_2}} = \frac{\epsilon_1}{\epsilon_2}$$

As regards the normal components, consider the pillbox gaussian surface shown; apply Gauss's theorem to it

$$D_{n_1} \Delta S - D_{n_2} \Delta S = \Delta Q = \rho_S \Delta S$$

and thus

$$D_{n_1} - D_{n_2} = \rho_S$$

But the electric flux density D can arise only as a result of *free* charges at the surface and none can arise in normal dielectrics. So taking $\rho_S = 0$ finally

$$D_{n_1} = D_{n_2}$$

and

$$\frac{E_{n_1}}{E_{n_2}} = \frac{\epsilon_2}{\epsilon_1}$$

Note that each is the reciprocal of the corresponding relation between the tangential components. Here it is the normal D field which is continuous and the normal E field which is discontinuous. Summarising for the dielectric to dielectric interface

E_t continuous, E_n discontinuous

D_t discontinuous, D_n continuous

These relations enable us to find expressions for the 'refraction' of E and D in passing from one dielectric to another. Suppose in *isotropic* dielectrics we consider the D and E fields separately as shown. Arbitrarily assume $\epsilon_1 > \epsilon_2$. Suppose

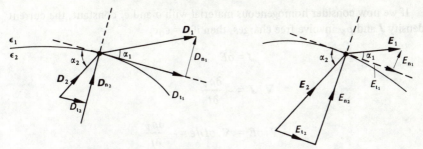

Figure 3.11   *Fields at the interface between dielectrics.*

we know ϵ_1, ϵ_2 and the incident conditions D_2, E_2, α_2; we wish to find D_1, E_1 and α_1. Refer to figure 3.11.

$$D_1 \sin \alpha_1 = D_2 \sin \alpha_2$$

$$D_1 \cos \alpha_1 = \frac{\epsilon_1}{\epsilon_2} D_2 \cos \alpha_2$$

$$D_1 = D_2 \left[\left(\frac{\epsilon_1}{\epsilon_2} \right)^2 \cos^2 \alpha_2 + \sin^2 \alpha_2 \right]^{1/2}$$

$$E_1 \cos \alpha_1 = E_2 \cos \alpha_2$$

$$E_1 \sin \alpha_1 = \frac{\epsilon_2}{\epsilon_1} E_2 \sin \alpha_2$$

$$E_1 = E_2 \left[\cos^2 \alpha_2 + \left(\frac{\epsilon_2}{\epsilon_1} \right)^2 \sin^2 \alpha_2 \right]^{1/2}$$

and from both sets of equations

$$\tan \alpha_1 = \frac{\epsilon_2}{\epsilon_1} \tan \alpha_2$$

Thus in passing from a low- to a high-permittivity region, D and E are refracted away from the normal and towards the tangent to the interface which may be convex or concave.

3.5.5   *Dielectric–conductor boundary; relaxation time*

Inside the conductor both electric displacement D and field strength E are zero and by considering the loop integral $\oint E \cdot dl = 0$ we see that there are no tangential components. The gaussian 'pillbox' shows that outside the conductor

$$|D| = |D_n| = |\epsilon E_n| = \rho_S$$

at the surface.

If we now consider homogeneous material with σ and ϵ_r constant, the current density J and ρ_V involve free charges, then if $\epsilon = \epsilon_0 \epsilon_r$

$$J = \sigma E$$

$$\nabla . J = -\frac{\partial \rho_V}{\partial t}$$

$$\nabla . \sigma E = \nabla . \sigma D / \epsilon = -\frac{\partial \rho_V}{\partial t}$$

Hence

$$\nabla . D = -\frac{\epsilon}{\sigma}\frac{\partial \rho_V}{\partial t}$$

Now using Maxwell's first equation

$$\operatorname{div} D = \rho_V = \nabla . D = -\frac{\epsilon}{\sigma}\frac{\partial \rho_V}{\partial t}$$

This gives a differential equation for ρ_V in which the variables can be separated

$$\frac{\partial \rho_V}{\rho_V} = -\frac{\sigma}{\epsilon}\partial t$$

$$\ln \rho_V = -\frac{\sigma}{\epsilon}t + \text{constant}$$

that is

$$\rho_V = \exp\left(-\frac{\sigma}{\epsilon}t + C\right)$$

Choose $\rho_V = \rho_{V0}$ at $t = 0$ and then

$$\rho_V = \rho_{V0}\exp\left(-\frac{\sigma}{\epsilon}t\right) = \rho_{V0}\exp\left(-\frac{t}{\tau}\right)$$

That is there is a decay of charge density ρ_V which is exponential with time. The time constant ϵ/σ is in seconds, and is called the *relaxation time*; it is noted that τ is short for good conductors, and long for good dielectrics.

For copper, conductivity $\sigma = 5.8 \times 10^7$ S/m and $\epsilon_r \approx 1$ so that relaxation time

$$\tau = \frac{\epsilon}{\sigma} = \frac{\epsilon_r \epsilon_0}{\sigma} = \frac{1}{36\pi 10^9}\frac{1}{5.8 \times 10^7}$$

or 1.53×10^{-19} s. On the other hand for quartz, which has conductivity σ of about 10^{-17} S/m and a dielectric constant ϵ_r of about 4, the relaxation time τ is

$$4\frac{1}{36\pi 10^9}\frac{1}{10^{-17}} = 3.5 \times 10^6 \text{ s}$$

Any charge density within a conductor decays exponentially with a time constant equal to the relaxation time. However, a point inside the conductor whose charge density is zero cannot become charged, so no decaying charge at one point can re-appear at any other. Nevertheless charge must be conserved, so any decaying charge must be balanced by the appearance of charge at the surface of the conductor. In conductors the relaxation time is so short that the decay of charge and the appearance of surface charge may be regarded as instantaneous.

3.6 Capacitors and capacitance

In its simplest form a capacitor has two conducting 'plates' with a dielectric between; this has the property of 'capacitance' C. Consider two arbitrarily shaped conductors in a homogeneous dielectric as illustrated in figure 3.12, conductor 1 carrying positive charge Q_+ and conductor 2 a negative charge Q_-. The charges are numerically equal and the conductors are isolated from other bodies. Conductor 1 is at a higher potential than 2, that is at a more positive potential, and work would be done in moving positive charge from 2 to 1. The potential difference V between the conductors is

$$V = V_1 - V_2 = -\int_2^1 E \,.\, \mathrm{d}l$$

where E is the electric field strength.

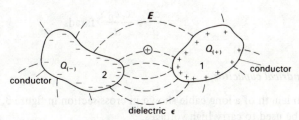

Figure 3.12 *A capacitor.*

By definition the capacitance between the two bodies (or 'plates') is the ratio between the magnitude of the charge on either conductor to the potential difference between them, that is

$$C = \frac{Q}{V} \text{ coulomb per volt (or farad)}$$

3.6.1 *Parallel-plate capacitor*

A parallel-plate capacitor is shown in figure 3.13. Suppose the plates shown form part of an extensive pair, which means no flux fringing.

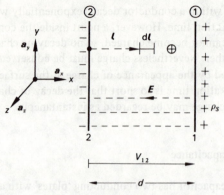

Figure 3.13   *Parallel-plate capacitor.*

$$E = -\frac{\rho_S}{\epsilon}a_x \qquad D = \epsilon E = -\rho_S a_x$$

$$V_{12} = -\int_2^1 E \cdot dl = -\int_2^1 \left(-\frac{\rho_S}{\epsilon}a_x\right) \cdot dl\,a_x = +\frac{\rho_S}{\epsilon}d = V \text{ volts}$$

$$Q = \rho_S S \qquad S = \text{plate area}$$

$$C = \frac{Q}{V} = \frac{\rho_S S}{\rho_S d/\epsilon}$$

$$= \frac{\epsilon S}{d} = \frac{\epsilon_r \epsilon_0 S}{d} \text{ farads}$$

3.6.2   Cylindrical capacitor

Consider 1 m length of a long cable shown in cross-section in figure 3.14; such a cable might be used to carry high voltages.

$$Q = \text{charge/m} = \text{C/m}$$

$$\rho_S = \frac{Q}{2\pi r} \text{ C/m}^2 = D$$

$$E = D/\epsilon = \frac{Q}{2\pi r\epsilon}a_r$$

$$dl = dr = dra_r$$

$$V_{21} = -\int_1^2 E \cdot dl = -\int_1^2 \frac{Q\,dr}{2\pi r\epsilon} = -\frac{Q}{2\pi\epsilon}\ln\left(\frac{r_2}{r_1}\right)$$

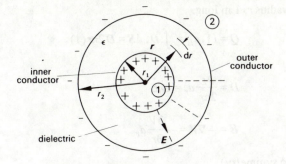

Figure 3.14 *Coaxial cylindrical capacitor.*

The inner surface, radius r_1, is at the higher potential so that

$$V_{12} = -V_{21} = \frac{Q}{2\pi\epsilon} \ln\left(\frac{r_2}{r_1}\right) = V$$

From which

$$C = \frac{Q}{V} = \left[\frac{2\pi\epsilon}{\ln(r_2/r_1)}\right] \text{ F/m}$$

For a length of x metres, $C = x$ [ ].

3.6.3 *Pair of parallel cylinders*

This is a very important case, for example the capacitance between two high-voltage transmission conductors. There are various approximations, but we shall develop an accurate solution. We need first an expression for the potential at a point P due to an infinite line charge ρ_L; see figure 3.15. The system has cylindrical symmetry; for ease instead of ρ for radius we use r. Consider the gaussian surface consisting

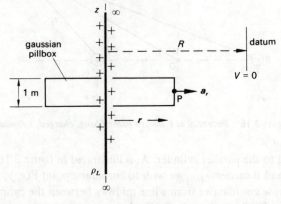

Figure 3.15 *Potential at a point P due to a long line charge ρ_L.*

of a cylinder radius r, 1 m long.

$$Q = (1)\rho_L = \int_S D \cdot dS = D\,2\pi r(1)$$

$$D = \frac{\rho_L}{2\pi r}\,a_r = \epsilon_0 E$$

$$E = -\nabla V = -\frac{\partial V}{\partial r}\,a_r$$

(It has z and ϕ symmetry)

$$dV = -E \cdot dr = -Edr = -\frac{\rho_L}{2\pi\epsilon_0}\frac{1}{r}\,dr$$

Hence

$$\text{work in moving } +1 \text{ C from } r \rightarrow R = -\frac{\rho_L}{2\pi\epsilon_0}\ln\left(\frac{R}{r}\right)$$

and then

$$\text{work in moving } +1 \text{ C from } R \rightarrow r = \frac{\rho_L}{2\pi\epsilon_0}\ln\left(\frac{R}{r}\right)$$

$$V_r = \frac{\rho_L}{2\pi\epsilon_0}(A - \ln r)$$

In this $A = \ln R$, and R is chosen at some convenient datum point for zero potential.

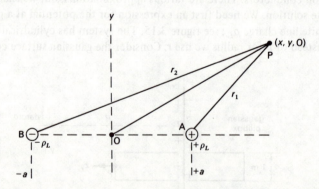

Figure 3.16   *Potential at P due to pair of long, charged, cylinders.*

Returning to the parallel cylinders A, B illustrated in figure 3.16, where A carries $+\rho_L$ and B carries $-\rho_L$, we wish to consider a point P(x, y, 0), in reality a line, and choose coordinates from a line midway between the cylinders; choose references $V = 0$ datum when $A = \ln R = \ln a$.

Potential at point P (really a line)

$$= \frac{\rho_L}{2\pi\epsilon_0} \left[\ln\left(\frac{a}{r_1}\right) - \ln\left(\frac{a}{r_2}\right) \right]$$

$$= \frac{\rho_L}{2\pi\epsilon_0} \ln\left(\frac{r_2}{r_1}\right) = \frac{\rho_L}{2\pi\epsilon_0} \frac{1}{2} \ln\left[\frac{(x+a)^2 + y^2}{(x-a)^2 + y^2}\right]$$

We cannot yet proceed to find the potential difference between the lines, for $x = a$ gives unmeaningful results; that is due to the practical impossibility of a line (that is no dimension) charge.

Suppose instead that we proceed to locate the equipotential surface of V_1 volts.

$$\frac{4\pi\epsilon_0 V_1}{\rho_L} = \ln [\quad] = \ln(K_1)$$

or

$$K_1 = \exp \frac{4\pi\epsilon_0 V_1}{\rho_L}$$

Hence for such a surface $K_1[(x-a)^2 + y^2] = (x+a)^2 + y^2$ and this is clearly a cylindrical surface. Rearranging

$$x^2 - 2ax\frac{K_1 + 1}{K_1 - 1} + y^2 + a^2 = 0$$

that is

$$\left[x - a\left(\frac{K_1 + 1}{K_1 - 1}\right)\right]^2 + y^2 = a^2 \frac{4K_1}{(K_1 - 1)^2}$$

Now the equation of a circle, centre $(c, 0)$ and radius b is

$$(x - c)^2 + y^2 = b^2$$

Hence an equipotential surface V_1 is a parallel cylinder centred at

$$c = \frac{a(K_1 + 1)}{K_1 - 1}$$

and of radius

$$b = \frac{a\,2(K_1)^{1/2}}{K_1 - 1}$$

where

$$K_1 = \exp\left(\frac{4\pi\epsilon_0 V_1}{\rho_L}\right)$$

The problem can now be reconsidered as a conductor of radius b, centre c away from a parallel earthed plane, as is shown in figure 3.17. We need the capacitance per metre from this to the earth. Its potential must be V_1.

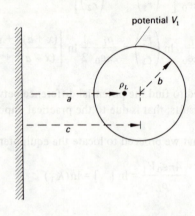

potential V_1

Figure 3.17   *Equipotential surface V_1 due to a pair of charged cylinders. a, distance from earthed plane where ρ_L is placed for equivalence. b, radius of V_1 surface. c, centre of V_1 surface. $V_1 = (\rho_L/4\pi\epsilon_0) \ln K_1 = (\rho_L/2\pi\epsilon_0) \ln (K_1)^{1/2}$.*

Usually we will know b and c but not a, and $c > a > b$. Thus eliminating a from the b and c equations produces

$$c^2 - b^2 = a^2 \frac{(K_1 + 1)^2 - 4K_1}{(K_1 - 1)^2} = a^2$$

$$a = (c^2 - b^2)^{1/2}$$

$$\frac{c}{a} = \frac{K_1 + 1}{K_1 - 1} = 1 + \frac{2}{K_1 - 1}$$

from which

$$\frac{1}{K_1 - 1} = \frac{c - a}{2a}$$

$$(K_1)^{1/2} = \frac{b}{2a}(K_1 - 1) = \frac{b}{2a}\frac{2a}{c - a} = \frac{b}{c - a} = \frac{b(c + a)}{c^2 - a^2} = \frac{c + a}{b}$$

Hence

$$a = (c^2 - b^2)^{1/2}$$

and

$$(K_1)^{1/2} = \frac{b}{c - a} = \frac{c + a}{b} = \frac{c + (c^2 - b^2)^{1/2}}{b}$$

K_1 and $(K_1)^{1/2}$ can now be computed in terms of the known b, c; proceeding

$$C = \frac{\rho_L}{V_1} = \frac{2\pi\epsilon_0}{\ln(K_1)^{1/2}} \text{ F/m to earth}$$

This can now be applied to the pair of cylinders; see figure 3.18.

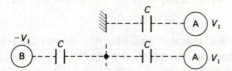

Figure 3.18 *Capacitance of a pair of parallel cylinders.*

Finally the results may be summarised

$$\text{cylinder to plane} \begin{cases} \dfrac{2\pi\epsilon_0}{\ln(K_1)^{1/2}} \text{ F/m} \\[3mm] \dfrac{2\pi\epsilon_0}{\cosh^{-1}(c/b)} \text{ F/m} \end{cases}$$

$$\text{cylinder to cylinder} \; \frac{\pi\epsilon_0}{\cosh^{-1}(c/b)} \text{ F/m}$$

The usual approximate results occur by taking $c = a$ and radius b as small, giving $(\pi/\epsilon_0)/\ln(2a/b)$.

Problems

3.1 Electrons are liberated at a long cylindrical wire surface *in vacuo* at N per second per metre length of wire. Assuming they disperse radially outwards with cylindrical symmetry according to

$$\rho = at^{3/2}$$

that there is no charge accumulation in the wire and that steady-state conditions exist, find expressions for D, ρ_V, J at the cylindrical surface $\rho = 2a$.

$$((A/a)(2)^{-1/3}a_\rho, \quad (A/a^2)(2)^{-1/3}/3, \quad (A/2a)a_\rho, \quad A = Nq_e/2\pi)$$

3.2 A dielectric disc, radius a, thickness d, is made into an electret by permanently polarising it throughout at P parallel to its axis. Find an expression for the electric potential V and field strength E on the axis, distant z from its surface, and approximate this for $z \gg d$.

$$\left(V = \frac{Pd}{2\epsilon_0} \left(1 - \frac{z}{(z^2 + a^2)^{1/2}} \right), \quad E = \frac{Pd}{2\epsilon_0} \frac{a^2}{(a^2 + z^2)^{3/2}} a_z \right)$$

3.3 Two extensive homogeneous isotropic dielectrics meet at the plane $z = 0$. Below the plane ($z < 0$) ϵ_{r_1} = 5, above ϵ_{r_2} = 2. A uniform electric field $E_1 = 2a_x - 3a_y + 5a_z$ kV/cm exists below the interface.
(a) Find E_2 above the boundary plane.
(b) Determine the energy densities in J/m^3 in both dielectrics.

((a) $2a_x - 3a_y + 12.5a_z$ kV/cm;   (b) 8·4 J/m^3, 15·0 J/m^3)

3.4 In an extensive volume of soil, liberated charges disperse freely from their centre of generation in spherical symmetry. Given that for soil ϵ_r = 3, $\sigma = 10^{-5}$ and that, when $t = 0$, $\rho_{V0} = 10^{-4}$ C/m^3 within the surface $r = 2$ m, find the current density and the current through the surface at times $t = 0$ and 10^{-5} s.

(25·13, 1263·3; 0·579, 29·12)

3.5 An electric field has potential $V = a\phi$ volts. Two long thin flat conducting plates, widths b, lying in the planes $\phi = 0$ and $\phi = \phi_2$ have their near edges in the lines $(\rho_1, 0, z)$ and (ρ_1, ϕ_2, z) in cylindrical coordinates.
(a) Find an expression for their capacitance per metre.
(b) Deduce an expression for a pair of parallel plates width b and separation d.

((a) $(\epsilon_0/\phi_2) \ln (\rho_1 + b/\rho_1)$;   (b) $\epsilon_0 b/d$)

3.6 An isotropic dielectric is placed in a uniform electric field such that a polarisation P exists within the dielectric. Show that the potential at the origin corresponds to an apparent surface charge density equal to the component of P normal to the surface plus an apparent volume charge density equal to $-$div P.

3.7 A parallel-plate capacitor has a spacing $2d$ and an area S. The space between the plates is filled with two dielectrics of dielectric constant ϵ_{r_1} and ϵ_{r_2}, conductivities σ_1 and σ_2. Each is of thickness d. A potential V is applied to the plate. Show that the surface charge density ρ_S between the dielectrics obeys the equation

$$\rho_S = \frac{V}{d} \frac{\epsilon_{r_2}\sigma_1 - \epsilon_{r_1}\sigma_2}{\sigma_1 + \sigma_2} \left[1 - \exp\left(-\frac{\sigma_1 + \sigma_2}{\epsilon_{r_1} + \epsilon_{r_2}} t \right) \right]$$

3.8 A conducting sphere of radius a carries a charge Q and is halfway submerged in a non-conducting liquid of dielectric constant ϵ_r. Calculate the surface charge density on (a) the sphere–air interface, and (b) the sphere–liquid interface.

3.9   A parallel-plate capacitor is totally immersed in a liquid of dielectric constant ϵ_r. Show that the difference in pressure between a point midway between the plates and a point on the median line outside the plates is

$$\frac{\epsilon_0(\epsilon_r - 1)E^2}{2}$$

where E is the electric field strength between the plates.

3.10   An infinite plane slab of homogeneous dielectric material of electric susceptibility χ_e lies in the $x = 0$ plane. A uniform time-varying electric field of strength $E = E_0 \cos(\omega t)a_x$ is applied to the slab. Show that the polarisation current density is

$$\frac{\epsilon_0 \chi_e E_0 \sin(\omega t)a_x}{1 + \chi_e}$$

4

The Magnetic Field

In chapter 2 we discussed the electric field arising from fixed or static charges and developed the concept of a force field in which such charges experience a force. Here we shall be concerned with the static magnetic field which arises from charges in motion. The motion of charges is equivalent to current flow which may arise from different current mechanisms. These include magnetisation currents in magnetic media, electron-beam currents in vacuum tubes and conduction currents in materials, all of which stem from the movement of charge. It makes little difference therefore if one considers the field arising from a permanent magnet or from a current-carrying coil; the basic effect, movement of charge, is the same. The approach used in this chapter will be to consider the effects due to coils carrying direct current.

4.1 Laws of magnetic force

The basic laws of magnetic force are based on a classical series of experiments first conducted by Oersted in Denmark and later extended by Ampère and his colleagues Biot and Savart in France.

4.1.1 Oersted's and Ampère's experiments

These experiments are concerned with the mechanical force F produced between two circuits carrying currents I_1 and I_2 as indicated in figure 4.1. The experiments established that the mechanical force F produced between two circuits can be represented by

$$F_{21} = k \oint_2 \oint_1 \frac{(I_2 \, dl_2) \times (I_1 \, dl_1 \times a_r)}{r^2}$$

66

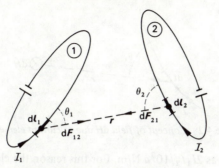

Figure 4.1 *Mechanical force F between two current circuits.*

In SI units the constant k is equal to $\mu_0/4\pi$ and the force is in newtons. μ_0 is the permeability of free space and is equal to $4\pi \times 10^{-7}$ henry per metre. The symbols $\oint_2$ and $\oint_1$ indicate that the integrals are performed over the closed circuits 2 and 1 respectively.

From the lack of symmetry in the expression it appears that $F_{12} \neq -F_{21}$; but it is possible to show that despite this lack of symmetry $F_{12} = -F_{21}$. The double loop integral is very difficult to perform and a field concept is better introduced.

Rewrite the equation

$$F_{21} = \oint_2 I_2 \; \mathrm{d}l_2 \times \frac{\mu_0}{4\pi} \oint_1 \frac{I_1 \; \mathrm{d}l_1 \times a_r}{r^2} \text{ newtons}$$

and introduce the field vectors

$$H = \oint_1 \frac{I_1 \; \mathrm{d}l_1 \times a_r}{4\pi r^2} \quad \text{and} \quad B = \mu_0 H$$

in free space. The magnetic field intensity or strength H is the magnetic force at element 2 due to the whole of circuit 1, and its units are amperes per metre. The magnetic induction, or flux density, B is measured in webers per square metre which for very little reason are now called tesla in SI units.

$F = \oint I \; \mathrm{d}l \times B$ newtons is the total force F experienced by the circuit 2 in a field of magnetic flux density B tesla.

4.1.2 Biot and Savart law

Of course the above equations are verifiable experimentally only in the integral or *closed circuit* form, for the current elements $I \; \mathrm{d}l$ are not physically realisable. However, using the element form indicated in figure 4.2

$$\mathrm{d}H = \frac{I \; \mathrm{d}l \times a_r}{4\pi r^2} \qquad\qquad \text{Biot and Savart law}$$

Biot and Savart (two colleagues of Ampère) actually showed that the mechanical force between two long parallel current conductors, distance a apart, could be

Figure 4.2   *Concept of field* dH *due to current element* I dl.

correctly computed as $2I_1I_2/10^7a$ N/m. For this reason the element form for the magnetic field is usually called the Biot and Savart law. But note it is also referred to as Ampère's rule, Laplace's rule, etc. It is *not* the only form which gives the correct results for the complete circuit.

4.1.3   Surface and volume elements

These are needed for sheet and volume current flow and are shown in figure 4.3. From figure 4.3a

$$I \, dl = K(dl \, db) = K \, dS$$

and from figure 4.3b

$$I \, dl = J(dy \, dz) \, dx = J \, d(vol)$$

Hence

$$H = \oint_l \frac{I \, dl \times a_r}{4\pi r^2} \qquad \text{line}$$

$$= \int_S \frac{K \, dS \times a_r}{4\pi r^2} \qquad \text{surface}$$

$$= \int_{vol} \frac{J \, d(vol) a_r}{4\pi r^2} \qquad \text{volume}$$

Note that the directional property is in l, K and J; it is not in I, as might be expected.

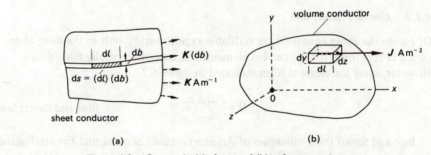

Figure 4.3   *Current in (a) sheet and (b) volume conductors.*

4.1.4 Field of straight conductor

This is a most useful form. With it we can retrace the Biot and Savart computation. Consider wire OE carrying current I along the z axis as is indicated in figure 4.4.

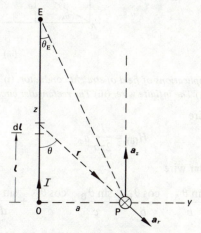

Figure 4.4   *Field due to a straight conductor.*

We wish to compute the magnetic field strength H at the point P, distance a from the end of the wire.

$$dH = \frac{I\, dl \times a_r}{4\pi r^2}$$

$$l = a \cot \theta; \qquad r = a \operatorname{cosec} \theta; \qquad dl = -a \operatorname{cosec}^2 \theta\, d\theta\, a_z$$

$$dH = \frac{I(-a \operatorname{cosec}^2 \theta\, d\theta)a_z \times a_r}{4\pi a^2 \operatorname{cosec}^2 \theta}$$

$$a_z \times a_r = a_\phi \sin \theta$$

$$dH = \frac{I}{4\pi a}(-\sin \theta\, d\theta)a_\phi$$

$$H = \frac{I}{4\pi a}\cos \theta_E a_\phi$$

This can be used for all circuits having straight-conductor sides. Note that mixed coordinates have been used above.

Three typical applications are given in figure 4.5.

   (i) The semi-infinite wire

$$H_{(i)} = \frac{I}{4\pi a}a_\phi$$

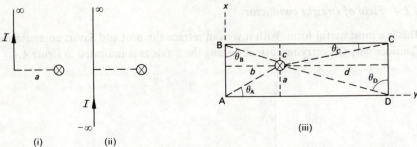

(i)   (ii)   (iii)

Figure 4.5   *Three applications of field of straight conductor.* (i) *The semi-infinite wire.*
(ii) *The infinite wire.* (iii) *The rectangular circuit.*

(ii) The infinite wire

$$H_{(ii)} = \frac{I}{2\pi a} a_\phi$$

(iii) The rectangular wire

$$H_{(iii)} = \frac{I}{4\pi}\left(\frac{\cos\theta_A}{a} + \frac{\sin\theta_A}{b} + \frac{\cos\theta_B}{b} + \frac{\sin\theta_B}{c} + \frac{\cos\theta_C}{c} + \frac{\sin\theta_C}{d} + \frac{\cos\theta_D}{d} + \frac{\sin\theta_D}{a}\right)a_z$$

Let us return now to the Biot and Savart calculation of the force between two
infinite wires illustrated in figure 4.6.

$$H = \frac{I_1}{2\pi a} a_\phi; \qquad B = \mu_0 H$$

$$F = I_2 l_2 \times B = I_2 l_2 \mu_0 \frac{I_1}{2\pi a} a_z \times a_\phi = I_1 I_2 l_2 \frac{4\pi}{10^7} \frac{1}{2\pi a}(-a_\rho)$$

$$\frac{F}{l_2} = \frac{2}{10^7} \frac{I_1 I_2}{a}(-a_\rho)\ \text{N/m}^{-1}$$

It is from this computation that the formal SI definition of the ampere is derived; that is currents of one ampere in conductors 1 m apart produce a force of
2×10^{-7} N m^{-1}.

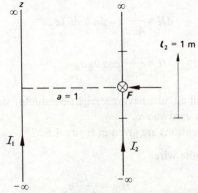

Figure 4.6   *Biot–Savart calculation; SI definition of 'ampere'.*

4.1.5 Ampère's circuital law

Consider a long straight wire carrying current I on the z axis, which is illustrated in figure 4.7. We already have by the Biot–Savart law

$$H = \frac{I}{2\pi a} a_\phi; \qquad dl = a \, d\phi \, a_\phi$$

$$\oint H \cdot dl = \int_0^{2\pi} \frac{I}{2\pi a} a_\phi \cdot a \, d\phi \, a_\phi$$

$$\oint H \cdot dl = \frac{I}{2\pi} \int_0^{2\pi} d\phi = I \qquad \text{Ampère's circuital law}$$

Although this integration path was taken as a concentric circle, the result would be the same for any closed path round the current I. Thus Ampère's circuital law states that, whatever the path, $\oint H \cdot dl =$ current enclosed (here I). In an analogous fashion to Gauss's law, it is extremely useful for application to circuits which have symmetry.

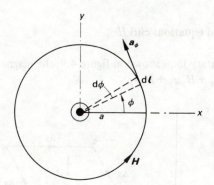

Figure 4.7   *Ampère's circuital law.*

Note particularly that because $\oint H \cdot dl = I \neq 0$ the magnetic field has curl and is therefore non-conservative.

4.1.6   Magnetic field near a current sheet

This is a simple application of Ampère's circuital law. Let the current sheet be in the $z = 0$ plane and $K = K_y a_y$ A/m as in figure 4.8a.

Suppose it is a very extensive sheet. Consider it to be composed of many parallel filaments. Clearly $H_y = 0 = H_z$; only H_x exists and it cannot vary spatially; for the rectangular loop

$$H_{12}l + H_{23}d + H_{34}l + H_{41}d = K_y l$$

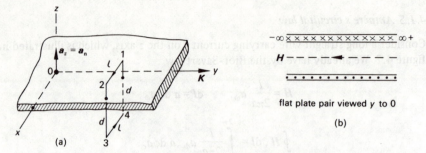

Figure 4.8   (a) *Single current sheet*. (b) *Flat plate pair*.

Since $H_z = -H_{23} = H_{41} = 0$ and $H_{12} = +H_{34} = H$ we have finally

$$H = \tfrac{1}{2}K_y a_x = \tfrac{1}{2}K_y a_y \times a_z = \tfrac{1}{2}K \times a_n$$

For the *flat plate pair* shown in figure 4.8b

$$H = K \times a_n \text{ inside}, \qquad H = 0 \text{ outside}$$

4.2   Maxwell's third equation: curl H

Consider the elementary loop shown in figure 4.9; the magnetic field strength H
at the centre is $H_x a_x + H_y a_y + H_z a_z$.

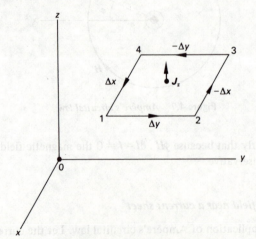

Figure 4.9   *Illustrating curl $H = J$*.

By definition

$$(\text{curl } H)_n = \lim_{\Delta S \to 0} \left(\frac{\oint H \cdot dl}{\Delta S} \right) a_n$$

ΔS is the surface of the small loop around which $H \cdot dl$ is to be summed. $\oint H \cdot dl / \Delta S$ is on a per unit area basis. Taking ΔS as $(\Delta x, \Delta y)$, that is parallel to the $z = 0$ plane

$$\oint H \cdot dl = \left[\left(H_y + \frac{1}{2} \frac{\partial H_y}{\partial x} \Delta x \right) \Delta y - \left(H_x + \frac{1}{2} \frac{\partial H_x}{\partial y} \Delta y \right) \Delta x \right.$$

$$\left. - \left(H_y - \frac{1}{2} \frac{\partial H_y}{\partial x} \Delta x \right) \Delta y + \left(H_x - \frac{1}{2} \frac{\partial H_x}{\partial y} \Delta y \right) \Delta x \right]$$

$$= J_z \, \Delta x \, \Delta y$$

$$(\text{curl } H)_z = \lim_{\Delta S \to 0} \left(\frac{H \cdot dl}{\Delta S} \right) a_z = \left(\frac{\partial H_y}{\partial x} - \frac{\partial H_x}{\partial y} \right) a_z = J_z a_z$$

There will be two similar expressions for the $(\text{curl } H)_x$ and $(\text{curl } H)_y$ components and the three need to be summed for the total curl H.

Summing for the three directions

$$\text{curl } H = \left(\frac{\partial H_y}{\partial x} - \frac{\partial H_x}{\partial y} \right) a_z + \left(\frac{\partial H_z}{\partial y} - \frac{\partial H_y}{\partial z} \right) a_x + \left(\frac{\partial H_x}{\partial z} - \frac{\partial H_z}{\partial x} \right) a_y$$

$$= \begin{vmatrix} a_x & a_y & a_z \\ \dfrac{\partial}{\partial x} & \dfrac{\partial}{\partial y} & \dfrac{\partial}{\partial z} \\ H_x & H_y & H_z \end{vmatrix}$$

$$= J_x a_x + J_y a_y + J_z a_z$$

$$\text{curl } H = \nabla \times H = J \qquad \text{Maxwell's third equation}$$

This is for steady fields and currents; for time-varying ones another term is added. This is essentially the point form of Ampère's circuital law, for the circuit or loop has been taken to its infinitesimal limit.

4.3 Maxwell's curl equations

Comparing the second and third equations

$$\text{curl } E = \nabla \times E = 0$$

$$\text{curl } H = \nabla \times H = J$$

there is obviously a difference between the electric field strength E and the magnetic field strength H fields (for non-time-varying quantities). It is that E has been derived from a conservative electric potential field of V, whereas H can be derived from a non-conservative magnetic vector potential field, A.

We now proceed to Stokes' theorem which is concerned with $\oint H \cdot dl$ taken over *large* surfaces of any shape, as distinct from the *point* aspect above.

4.4 Stokes' theorem

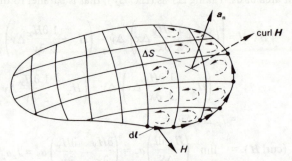

Figure 4.10 *Illustrating Stokes' theorem.*

This replaces a surface integral by a loop integral or vice versa. Consider any closed curve and any surface whose boundary is the curve. Break the surface into incrementals ΔS as we have done in figure 4.10. For each cell, $\oint H \cdot dl / \Delta S$ is the magnitude of the component of curl H normal to the cell surface, that is along a_n.

$$\frac{\oint_{\Delta S} H \cdot dl}{\Delta S} = (\nabla \times H) \cdot a_n$$

Rearranging and taking $(\Delta S)a_n = \Delta S$, the *circulation* round the cell is

$$\oint_{\Delta S} H \cdot dl = (\nabla \times H) \cdot a_n \, \Delta S = \nabla \times H \cdot \Delta S$$

The suffix ΔS indicates that the closed line integral is carried out around the periphery of ΔS. As we proceed thus round each of the incremental areas ΔS to cover the surface, cancellation of the contributions $\oint_{\Delta S} H \cdot dl$ takes place everywhere except at the boundary line, so that the $\oint H \cdot dl$ need be made only round the periphery of S. Proceeding to the limit with ΔS we have finally

$$\oint_S H \cdot dl = \oint_S (\nabla \times H) \cdot dS \qquad \text{Stokes' theorem}$$

The loop integral of the dot product $H \cdot dl$ taken round a boundary path of S is the same as the surface integral of the curl $H \ (\nabla \times H)$ taken over the surface area of S.

Stokes' theorem is an identity and it holds for a surface in any vector field.

4.4.1 Deduction of Ampère's circuital law

Start from Maxwell's third equation, relating curl H and J.

$$\nabla \times H = J$$

Hence

$$\int_S (\nabla \times H) \cdot dS = \int_S J \cdot dS = I$$

But from Stokes' theorem the left-hand side can be replaced by $\oint_S H \cdot dl$. Hence $\oint_S H \cdot dl = I$ which is Ampère's circuital law.

4.4.2 div curl F = 0

In section 1.6 it has already been stated that $\nabla \cdot \nabla \times F$ is zero for any vector field F. Now it can be proved by the use of Stokes' theorem.

Suppose div curl $F = \nabla \cdot \nabla \times F = T$ which must be scalar. We need to show $T = 0$

$$\int_{\text{vol}} \nabla \cdot (\nabla \times F) \, d(\text{vol}) = \int_{\text{vol}} T \, d(\text{vol})$$

Now employing the divergence theorem, replacing the volume integral by an equivalent surface integral

$$\int_{\text{vol}} \nabla \cdot (\nabla \times F) \, d(\text{vol}) = \oint_S (\nabla \times F) \cdot dS = \oint_S F \cdot dl$$

by Stokes' theorem. But this last is now an integral over a *closed* surface, which has no periphery l and the integral must be zero. Hence $\int_{\text{vol}} T \, d(\text{vol}) = 0$ and $T = 0$. Finally div curl $F = \nabla \cdot \nabla \times F = 0$.

4.4.3 Continuity equation

$$\nabla \cdot J = \text{div } J = -\frac{d}{\partial t}(\rho_V)$$

For non-time-varying fields, $d\rho_V/\partial t \to 0$ so that the continuity equation becomes $\nabla \cdot J = 0$.

Starting from Maxwell's third equation, $\nabla \times H = J$, take div and use the previous result

$$\text{div curl } H = \nabla \cdot \nabla \times H = \nabla \cdot J = 0$$

4.4.4 Example on curl F

Suppose we need to compute the circulation of F around a given path, as shown in figure 4.11.

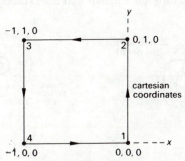

Figure 4.11 *Circulation of F around given closed path.*

$$F = 3 \cos(\pi x) \sin(\pi y)a_x - 4 \sin(\pi x) \cos(\pi y)a_y - 6 \cos(\pi x) \cos(\pi y)a_z$$

$$\int_S (\nabla \times F) \cdot dS = \int_S F \cdot dl \qquad \text{Stokes' theorem}$$

$$12 = \int F \cdot a_y \, dy = \int_0^1 -4 \sin(\pi x) \cos(\pi y) \, dy = 0 \qquad (x = 0)$$

$$23 = \int F \cdot a_x \, dx = \int_0^{-1} 3 \cos(\pi x) \sin(\pi y) \, dx = 0 \qquad (y = 1)$$

$$34 = \int F \cdot a_y \, dy = \int_1^0 -4 \sin(\pi x) \cos(\pi y) \, dy = 0 \qquad (x = -1)$$

$$41 = \int F \cdot a_x \, dx = \int_{-1}^0 3 \cos(\pi x) \sin(\pi y) \, dx = 0 \qquad (y = 0)$$

Hence there is no circulation and we can write $\int_S \nabla \times F \cdot dS = 0$ although this was computed by the line integral round the periphery using Stokes' theorem.

4.5   Maxwell's fourth equation: div B

We need to start by defining flux density and flux for the magnetic field

$$B = \mu_0 H \text{ Wb m}^{-2} \text{ (T)}; \qquad \mu_0 = 4\pi \times 10^{-7} \text{ H/m for vacuum}$$

$$\Phi = \int_S B \cdot dS \text{ Wb}$$

This is for an open surface, S, compared with electric flux

$$\Psi = \oint_S D \cdot dS = Q \text{ C}$$

for a closed surface. The difference here is that lines (tubes) of Ψ *must* start at a

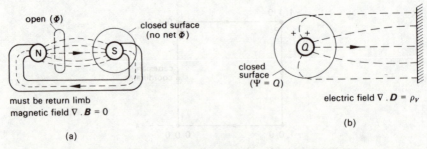

open (Φ)

closed surface
(no net Φ)

must be return limb
magnetic field $\nabla \cdot B = 0$

(a)

closed
surface
($\Psi = Q$)

electric field $\nabla \cdot D = \rho_V$

(b)

Figure 4.12   *Divergence in* (a) *magnetic and* (b) *electric fields.*

positive charge and finish at a negative charge; whereas lines of Φ are continuous and close on themselves. In fact there are *no isolated magnetic poles*.

Figure 4.12 indicates the difference between divergence in electric and magnetic fields.

We can now state Maxwell's fourth equation simply. For a closed surface

$$\oint_S B \cdot dS = 0 = \int_{vol} \nabla \cdot B \, d(vol)$$

Hence we can apply this to the volume element d(vol) itself and obtain $\nabla \cdot B \, d(vol) = 0$; finally the point form

$$\text{div } B = \nabla \cdot B = 0 \qquad \text{Maxwell's fourth equation}$$

4.6 Maxwell's equations for steady (d.c.) fields

They can be stated in differential and integral forms.

Differential (point) form | Integral form

$$\text{div } D = \nabla \cdot D = \rho_V \ (1) \qquad \oint_S D \cdot dS = Q = \int_{vol} \rho_V \, d(vol)$$

$$\text{curl } E = \nabla \times E = 0 \ (2) \qquad \oint E \cdot dl = 0$$

$$\text{curl } H = \nabla \times H = J \ (3) \qquad \oint H \cdot dl = I = \int_S J \cdot dS$$

$$\text{div } B = \nabla \cdot B = 0 \ (4) \qquad \oint_S B \cdot dS = 0$$

There is no generally accepted numbering system for these four equations; here the numbers have been chosen so that the first pair refer to the *electric* field and the second pair to the *magnetic* field. The choice of which of the two forms, differential or integral, to use will depend on the particular problem in hand. The divergence equations are consistent for static and time-varying fields but the curl equations are not and an extra term appears in the curl equations for time-varying fields.

4.7 Magnetic potential

The vector field E was derived from the scalar electric potential V by $E = -\text{grad } V = -\nabla V$. Some field problems were simplified by its use.

In a similar way it may help if we can derive the magnetic field strength H (and flux density B) from a magnetic potential field. This is not so easily realisable as that of the electric potential which arose from considering the work done in positioning a unit charge. In the magnetic potential case physically we have no unit magnetic source to help us, and the potential field is really contrived as an aid rather than a physical reality. We shall see that the magnetic potential can itself be considered either as a scalar U or as a vector A; both are important. The magnetic field strength H may be expressed as the gradient of U and the magnetic induction B may be considered as the curl of A.

4.7.1   Scalar magnetic potential U

Consider a coil to be rectangular and to carry a current I as shown in figure 4.13a. Let the coil suffer a bodily displacement dr. The effect of the displacement is viewed from P and the element $I\,dl$ leads to a projected area change as viewed from P of $(dr \times dl) . a_r$

$$\text{solid angle change} = \frac{(dr \times dl) . a_r}{r^2}$$

(see appendix 3). The convention regarding positive Ω is that, when viewed from P, the current is circulating anti-clockwise. In the elementary periphery it is clockwise so that the change of solid angle is $-(dr \times dl) . a_r/r^2$ due to dr, dl and the total change of solid angle

$$d\Omega = \oint \frac{-(dr \times dl) . a_r}{r^2} = \nabla\Omega . dr$$

$$= -\oint \frac{dl \times a_r}{r^2} . dr = -\frac{4\pi}{I} \oint \frac{I\,dl \times a_r}{4\pi r^2} . dr = -\frac{4\pi}{I} H . dr$$

Hence

$$H = -\frac{I}{4\pi}\nabla\Omega = -\nabla\frac{I\Omega}{4\pi} = -\nabla U = -\text{grad } U$$

Comparing this with the electric field strength $E = -\nabla V$, the quantity $U = I\Omega/4\pi$ is analogous to the scalar electric potential V; it is the *scalar magnetic potential* of a point at which a circuit subtends solid angle Ω.

Scalar magnetic potential U is classically derived by means of Ampère's equi-

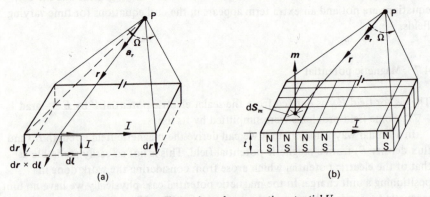

Figure 4.13   *Change in scalar magnetic potential U.*

valent magnetic shell; this is a surface of magnetic dipoles, with its periphery on the current-carrying circuit. Its magnetic moment is (pole strength)(t)(Wb m) and is directed normal to the magnetised surface. This concept is illustrated in figure 4.13b.

$$m = \text{(pole strength)}(t)a_n \text{ Wb m}$$

By analogy with the electric dipole already discussed for an element dm of area dS_m, we have for the magnetic potential dU at P

$$dU = \frac{dm \cdot (-a_r)}{4\pi\mu_0 r^2}$$

$$= \frac{I}{4\pi} d\Omega = \frac{I}{4\pi} \frac{dS_m \cdot (-a_r)}{r^2}$$

This result must hold for the magnetic shell to be equivalent to a current I round the periphery of dS_m. Finally for equivalence the elementary magnetic moment is

$$dm = (\mu_0 I) \, dS_m \text{ Wb m}$$

The potential U used to be associated with the work done in positioning a (fictitious) unit north pole at P.

The main difficulty with the scalar magnetic potential U is that, unlike V, it is *multi-valued*. For example, each time the unit pole is taken from P through the circuit back to P, the solid angle has increased by 4π so that $U = (I/4\pi)(\Omega \pm 4\pi \pm 8\pi,$ etc.). This change in U on linking the circuit has its counterpart in the energy change in the battery to keep the current I constant against induced voltages.

The magnetic shell acts as an artificial barrier limiting U to its principal value. But for current-carrying coils the multi-valued nature of U is real and cannot be so eliminated. In reality scalar magnetic potential is not a conservative field, borne out by the fact already established that $\oint H \cdot dl = I \neq 0$.

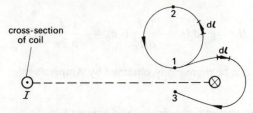

Figure 4.14 *Multi-valued nature of scalar U.*

A coil carrying a current I is shown in figure 4.14 and we consider certain closed paths near to the coil. Consider the line integral $\int H \cdot dl$ and path element dl. $U = f(xyz)$; $dU = \text{grad } U \cdot dl = -H \cdot dl$. For the closed path 1, 2, 1 $\oint H \cdot dl = -\int dU = -\int (I/4\pi) \, d\Omega = 0$ for there is in this case no change in Ω.

But for the path 1 to 3 we have $\int H \cdot dl = U_1 - U_3 \approx (I/4\pi) [2\pi - (-2\pi)]$; closing 3 to 1 we have finally

$$\oint H \cdot dl = I$$

which is Ampère's circuital law. Let us now apply the scalar magnetic potential U to one or two actual cases.

4.7.2   Infinite straight wire carrying steady current I

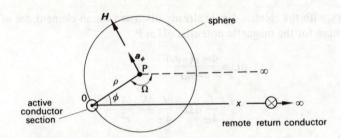

Figure 4.15   *Scalar magnetic potential of long straight conductor.*

A section of the conductor is shown in figure 4.15 and we assume the return conductor to be infinitely remote on the x axis side. Choose the current directions so that Ω as viewed from P is positive (current anti-clockwise).

$$\frac{\Omega}{4\pi} = \frac{\pi - \phi}{2\pi} \qquad \Omega = 2(\pi - \phi)$$

ρ, ϕ are cylindrical coordinates. At P, the magnetic scalar potential

$$U = \frac{I}{4\pi} \Omega = \frac{I}{2\pi} (\pi - \phi)$$

But

$$H = -\mathrm{grad}\, U = -\frac{1}{\rho} \frac{\partial}{\partial \phi} U\, a_\phi = -\frac{1}{\rho} \frac{I}{2\pi} (-1) a_\phi = \frac{I}{2\pi\rho} a_\phi$$

This of course can be more easily obtained by Ampère's rule.

4.7.3   Field on axis of a circular loop of current I

Consider a ring of radius a in the yz plane. The axis of the ring is parallel to the x axis and the current flow I is such as to give a positive value of Ω at P. This is illustrated in figure 4.16. The area of the cap defined by the coil on the spherical surface of radius r is $2\pi r^2 (1 - \cos \phi)$ and

$$\Omega_\mathrm{P} = 2\pi(1 - \cos \phi) = 2\pi \left[1 - \frac{x}{(x^2 + a^2)^{1/2}} \right]$$

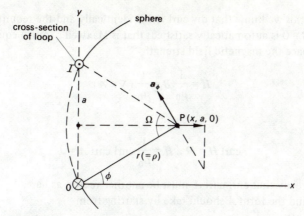

Figure 4.16 *Scalar magnetic potential of circular loop of current.*

Hence the magnetic field strength H_P at P is given by

$$H_P = -\text{grad } U = -\nabla \frac{I}{2} \left[1 - \frac{x}{(a^2 + x^2)^{1/2}} \right] = -\nabla \frac{I}{2} f(x)$$

$$= -\frac{I}{2} \frac{\partial f}{\partial x} a_x = +\frac{I}{2} [(a^2 + x^2)^{-1/2} - x^{1/2}(a^2 + x^2)^{-3/2} 2x] a_x$$

$$= \frac{I}{2} (a^2 + x^2)^{-3/2}(a^2 + x^2 - x^2) a_x = \frac{I}{2} \frac{a^2}{(a^2 + x^2)^{3/2}} a_x$$

$$H_P = H_x = \frac{I}{2a} \sin^3 \phi a_x$$

which is correct. Note, however, that difficulties arise if we work in cylindrical coordinates, replacing r by ρ. Taking $-\text{grad } U$ we obtain

$$H = -\frac{I}{2} \frac{1}{\rho} \sin \phi a_\phi$$

To obtain H_x take $a_\phi = -\sin \phi a_x$ and $\rho = a/\sin \phi$ to give

$$H_x = \frac{I}{2a} \sin^3 \phi a_x$$

4.8 Vector magnetic potential A

This is of value for inductance studies, for time-varying fields, etc. It is introduced as another aid in evaluating magnetic flux density B and field strength H fields. A can be linked directly with H or B; we choose B and define A so that $B = \text{curl } A = \nabla \times A$.

In doing this we know that div curl $A = 0$ identically and the requirement div $B = \nabla . B = 0$ is automatically satisfied; that is Maxwell's fourth equation.

In free space the magnetic field strength

$$H = \frac{1}{\mu_0} B = \frac{1}{\mu_0} (\nabla \times A)$$

and also

$$\text{curl } H = \nabla \times H = \frac{1}{\mu_0} \text{curl curl } A = J$$

We know already that curl curl A must be another vector, as here.

We can find the form A should take by starting from

$$B = \frac{\mu_0 I}{4\pi} \oint \frac{dl \times a_r}{r^2} = \frac{\mu_0 I}{4\pi} \oint \left[+ \nabla \left(\frac{1}{r} \right) \times dl \right]$$

in which $\nabla (1/r) = -(1/r^2) a_r$ has been used.

Now curl $SF = S$ curl $F +$ grad $S \times F$ (identically). Rearrange this as $\nabla \times SF = S\nabla \times F - F \times \nabla S$; substitute $S = 1/r$ and $F = dl$. The B relation becomes

$$B = \frac{\mu_0 I}{4\pi} \left[\oint \nabla \times \frac{dl}{r} - \oint \frac{1}{r} \nabla \times dl \right]$$

∇ is an operator in the space coordinates of r, not of dl, so that $\nabla \times dl = 0$ and the second integral is zero.

$$B = \frac{\mu_0 I}{4\pi} \oint \nabla \times \frac{dl}{r} = \frac{\mu_0 I}{4\pi} \nabla \times \oint \frac{dl}{r}$$

$$= \nabla \times \frac{\mu_0 I}{4\pi} \oint \frac{dl}{r} = \nabla \times A$$

Thus

$$A = \frac{\mu_0}{4\pi} \oint \frac{I \, dl}{r} \qquad \text{line current}$$

$$= \frac{\mu_0}{4\pi} \int_S \frac{K \, dS}{r} \qquad \text{surface current}$$

$$= \frac{\mu_0}{4\pi} \int_{vol} \frac{J \, d(vol)}{r} \qquad \text{volume current}$$

As before the zero reference for $A = 0$ is at infinity.

If we use the current-element form for dA we have respectively

$$\frac{\mu_0}{4\pi} \frac{I \, dl}{r}, \frac{\mu_0}{4\pi} \frac{K \, dS}{r}, \frac{\mu_0}{4\pi} \frac{J \, d(vol)}{r}$$

as with the current-element Biot and Savart relation, only the completed integral for the whole circuit has significance

$$dB = \nabla \times dA \qquad \text{and} \qquad dH = \frac{1}{\mu_0} \nabla \times dA$$

Now consider applications of the magnetic vector potential.

4.8.1 Laplacian of A, $\nabla^2 A$

We shall deduce this by analogy with the laplacian of V, the electric scalar potential, and we start from Maxwell's first equation $\nabla . D = \rho_V$.

$$\nabla . (\epsilon_0 E) = \rho_V; \qquad \nabla . E = \frac{\rho_V}{\epsilon_0}; \qquad \nabla . (-\nabla V) = \frac{\rho_V}{\epsilon_0}$$

so finally

$$\nabla . \nabla V = \nabla^2 V = -\frac{\rho_V}{\epsilon_0} \qquad \text{Poisson}$$

For a volume charge

$$V = \frac{1}{4\pi\epsilon_0} \int_{\text{vol}} \frac{\rho_V \, d(\text{vol})}{r}$$

and for this we know that

$$\nabla^2 V = \frac{-\rho_V}{\epsilon_0}$$

For A we have a defining relation

$$A = \frac{\mu_0}{4\pi} \int_{\text{vol}} \frac{J \, d(\text{vol})}{r}$$

Suppose we consider x, y, z components of A separately, that is

$$A_x = \frac{\mu_0}{4\pi} \int_{\text{vol}} \frac{J_x \, d(\text{vol})}{r}$$

this is now a scalar, analogous to V.

Hence

$$\nabla^2 A_x = -\mu_0 J_x; \qquad \nabla^2 A_y = -\mu_0 J_y; \qquad \nabla^2 A_z = -\mu_0 J_z$$

Thus

$$\nabla^2 A = \nabla^2 A_x a_x + \nabla^2 A_y a_y + \nabla^2 A_z a_z$$
$$= -\mu_0(J_x a_x + J_y a_y + J_z a_z) = -\mu_0 J$$

This is Poisson's equation for free space. In a medium which is homogeneous and isotropic $\mu = \mu_r\mu_0$, μ_r being the relative permeability.

$$\nabla^2 A = -\mu_r\mu_0 J = -\mu J \qquad\qquad \text{Poisson}$$

Clearly for a current-free region $J = 0$ and $\nabla^2 A = 0$, which is Laplace's equation again.

In the previous section it was shown that curl curl $A = \mu J$ and here that lap $A = -\mu J$. Hence since

$$\nabla \times \nabla \times A = \nabla\nabla . A - \nabla^2 A$$

$$\nabla\nabla . A = 0 \qquad \text{or} \qquad \text{div } A = 0$$

for the quasi-static case we have considered.

4.8.2   *Magnetic flux linking a circuit*

Consider a closed circuit in a magnetic field of density B as indicated in figure 4.17. We are interested in the magnetic flux Φ linking the circuit. By definition $B = \nabla \times A$ and, for the flux through a surface area element dS

$$d\Phi = B . dS = (\nabla \times A) . dS$$

which gives

$$\Phi = \int_S B . dS = \int_S (\nabla \times A) . dS$$

This is an integral taken over the surface S whose periphery is the circuit through which we need the flux Φ. Now use Stokes' theorem to obtain

$$\Phi = \int_S B . dS = \int_S (\nabla \times A) . dS = \oint_l A . dl$$

The latter is a loop integral round the circuit periphery. For a filament (which has no cross-sectional area) this is well defined; but even then there are difficulties.

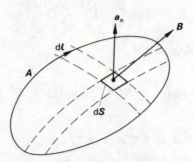

Figure 4.17   *Flux Φ linking a circuit.*

4.9   Inductance in terms of magnetic potential

The usual definition of inductance L is flux linkages per ampere or $L = \lambda/I$ for a linkage λ and current I. Here a difficulty arises in defining flux linkages. If there *are* N distinct turns which are themselves filaments, each having $\Phi_1, \Phi_2, \ldots, \Phi_N$ flux passing through them, $\lambda = \Phi_1 + \Phi_2 + \cdots + \Phi_N$.

In practice the situation is never so clear, for the turns are not distinct, for example, in an open helix, and the conductor is not a filament — it has a cross-section s.

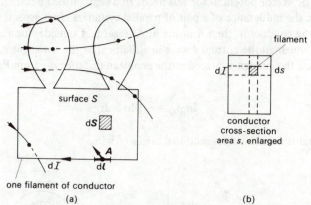

one filament of conductor

(a)   (b)

Figure 4.18   *Flux linkage with one filament of conductor.*

We now refer to figure 4.18a which shows flux linkage associated with one filament of a conductor shown in cross-section in figure 4.18b.

Suppose we break down the conductor into a large number of elementary 'filaments' which traverse the complete circuit, figure 4.18a. If the filament encloses a surface S which has no re-entrancy (the case of re-entrancy of the surface, for example, open helix, is fully discussed in Carter (1967)) the flux linkage through it is definite, including the turns

$$\Phi = \oint_S A \cdot dl = \oint_S A_l \, dl$$

In this A_l is the component of A along the length dl of the filament. But this filament is carrying only dI, not the full current I, so that

$$d\lambda = \oint_S A_l \, dl \, \frac{dI}{I}$$

This now must be summed (integrated) for all the filaments, that is over the (little) area s (see figure 4.18b)

$$\lambda = \int_s \oint_S A_l \, dl \, \frac{dI}{I}$$

$$L = \int_s \oint_S A_l \, dl \, \frac{dI}{I^2}$$

In this dI is being integrated over the conductor cross-section s and dl over the filamentary periphery bounding S.

If I, J and area s are constant throughout the conductor, $dI/I = ds/s$ and this then reduces to

$$L = \int_s \oint_S A_l \, dl \, \frac{ds}{Is} \quad \text{henrys}$$

4.9.1 Long cylindrical conductor

The magnetic vector potential for this needs to be established before proceeding to compute the inductance of a pair of parallel cylinders. Two aspects arise: first A outside the conductor, then A inside it. As regards A outside, for a cylinder it will be as though all the current I were in a filament at the wire centre. For any other section the current is placed at the geomean distance ρ_G from P, as shown in figure 4.19a

$$\ln \rho_G = \frac{1}{s} \int_s \ln \rho \, ds$$

This integral is over the cross-sectional area s of wire.

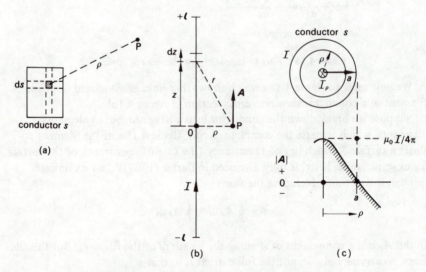

Figure 4.19   *Vector potential A for a long conductor.*

Outside the wire A_o, see figure 4.19b, refers to the magnetic vector potential outside the wire

$$A_o = \frac{\mu_0 I}{4\pi} \oint \frac{dz a_z}{r} = \frac{\mu_0 I}{4\pi} \int_{-l}^{+l} \frac{dz a_z}{(z^2 + \rho^2)^{1/2}}$$

$$A_o = \frac{\mu_0 I}{2\pi} \ln \left[\frac{l}{\rho} + \left\{ \left(\frac{l}{\rho} \right)^2 + 1 \right\}^{1/2} \right] a_z \rightarrow \frac{\mu_0 I}{2\pi} \ln \left(\frac{2l}{\rho} \right) a_z$$

For long lines $l/\rho > 1$.

$$A_\text{o} = \left(\frac{\mu_0 I}{2\pi} \ln 2l - \frac{\mu_0 I}{2\pi} \ln \rho \right) a_z$$

We can choose our reference $A = 0$ anywhere and write

$$A_\text{o} = \left(C_\text{o} - \frac{\mu_0 I}{2\pi} \ln \rho \right) a_z$$

Inside the wire A_i, see figure 4.19c, refers to the magnetic vector potential inside the wire. If the current density is uniform

$$I_\rho = I \frac{\rho^2}{a^2}$$

$$B = \frac{\mu_0 I}{2\pi\rho} \left(\frac{\rho}{a} \right)^2 a_\phi = \nabla \times A = -\left(\frac{\partial A_z}{\partial \rho} \right) a_\phi$$

(no other terms)

$$\frac{\partial A_z}{\partial \rho} = -\frac{\mu_0 I}{2\pi} \frac{\rho}{a^2}$$

$$A_\text{i} = A_z = \left[C_\text{i} - \frac{\mu_0 I}{4\pi} \left(\frac{\rho}{a} \right)^2 \right] a_z$$

When A_o and A_i are combined clearly we must choose C_o and C_i so that at the surface A has the same value. We do so by taking $A = 0$ at $\rho = a$. This gives

$$A_\text{o} = -\frac{\mu_0 I}{2\pi} \ln \left(\frac{\rho}{a} \right) a_z$$

$$A_\text{i} = +\frac{\mu_0 I}{4\pi} \left[1 - \left(\frac{\rho}{a} \right)^2 \right] a_z$$

The variation of the magnetic vector potential with distance is shown in figure 4.19c. It would of course have been possible to have taken A_o positive and A_i negative. Note also that if the conductor has a permeability μ_r then A_i, but not A_o, would have been increased. For an infinitely long conductor $\text{d}A$ is proportional to $\text{d}z/z$ for values of z where $r \gg \rho$. The rather unusual outcome of this is that $A \to \infty$ logarithmically, which at first sight is worrying, but not if one remembers that the derivative of A is *not* necessarily infinite. In calculations of potential, whether electric or magnetic, involving line distribution and cylindrical symmetry, the choice of zero of potential at infinity is unsuitable as this appears to give an infinite potential at all points!

The value of the magnetic flux density B obtained from curl A for $\rho > a$ is

$$B = \nabla \times A = \frac{\mu_0 I}{2\pi\rho} a_\phi$$

in agreement with the Biot–Savart law.

4.9.2 Inductance of a pair of long parallel cylinders

This is the counterpart of the capacitance calculation made in section 3.6.3 and has obvious application in the field of high-voltage transmission. The cross-section of the cylinders is shown in figure 4.20. Earlier in this chapter the expression for

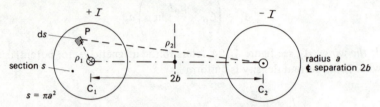

Figure 4.20 *Inductance of a pair of parallel cylinders.*

inductance in terms of magnetic vector potential was obtained. In the expression $\int_s \oint_S A_l \, dl \, ds/Is$, the filament has element length $dl = dz$ so that the loop integral $\oint_S$ is along the infinite wire and is the same for all filaments. This can be removed, and instead a length of 1 m only is considered.

$$L = \int_s \frac{A_l \, ds}{Is} \ \text{H m}^{-1}$$

The magnetic vector potential A at P has a contribution A_i from conductor C_1 and A_o from conductor C_2, so for conductor C_1

$$A = \left\{ \frac{\mu_0 I}{4\pi} \left[1 - \left(\frac{\rho_1}{a} \right)^2 \right] - \frac{\mu_0(-I)}{2\pi} \ln \left(\frac{\rho_2}{a} \right) \right\} a_z$$

In these $A_l = A_z$, so we can write, cancelling the currents I

$$Ls = \int_s \frac{\mu_0}{4\pi} \left[1 - \left(\frac{\rho_1}{a} \right)^2 \right] ds + \int_s \frac{\mu_0}{2\pi} \ln \left(\frac{\rho_2}{a} \right) ds$$

For the first of these integrals we have cylindrical symmetry so that we may use $ds = 2\pi\rho_1 \, d\rho_1$. In the second we can replace ρ_2 by $\rho_G = 2b$; this is a constant and the integrals reduce to give

$$Ls = \int_0^a \frac{\mu_0}{4\pi} \left[1 - \left(\frac{\rho_1}{a} \right)^2 \right] 2\pi\rho_1 \, d\rho_1 + \frac{\mu_0}{2\pi} \left[\ln \left(\frac{2b}{a} \right) \right] s$$

$$= \frac{\mu_0}{4\pi} \left| \tfrac{1}{2}\rho_1{}^2 - \frac{\rho_1{}^4}{4a^2} \right|_0^a (2\pi) + \frac{\mu_0}{2\pi} \left[\ln \left(\frac{2b}{a} \right) \right] s$$

where $s = \pi a^2$. Hence

$$L = \frac{\mu_0}{2\pi} \left[\tfrac{1}{4} + \ln \left(\frac{2b}{a} \right) \right] \text{ henrys per metre per wire}$$

There will of course be an exactly similar contribution for the second conductor C_2 making finally

$$L = \frac{\mu_0}{\pi} \left[\tfrac{1}{4} + \ln \left(\frac{2b}{a} \right) \right] \text{ henrys per metre of circuit}$$

The $\tfrac{1}{4} (\mu_0/\pi)$ term is that due to the internal flux linkages within the conductors. When $2b$ is large compared with a, then the internal linkage term can be omitted, giving $L = (\mu_0/\pi) \ln (2b/a)$ H/m, a result more readily obtained otherwise as follows.

4.9.3 Inductance of a pair of cylinders (alternative approach)

The two cylinders are shown in figure 4.21 with currents $+I$ in conductors C_1 and $-I$ in conductor C_2. One simple approach ignoring the internal flux linkages in

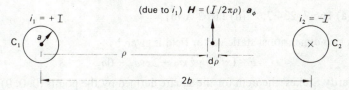

Figure 4.21 *Simple approach to inductance of a pair of parallel cylinders.*

the wires is shown. The field B between the conductors due to $i_1 = +I$ in C_1 is $(\mu_0 I/2\pi\rho)a_\phi$. The flux linkage contribution due to i_1 will be

$$\lambda = \int_a^{2b} \frac{\mu_0 I}{2\pi\rho} \, d\rho = \frac{\mu_0 I}{2\pi} \ln \left(\frac{2b}{a} \right)$$

In this i_2 has been assumed concentrated on the axis of C_2 and the integral carried out between there and the surface of C_1. There will, by symmetry, be an equal linkage contribution due to i_2. Finally, and as before

$$L = \frac{\mu_0}{\pi} \ln \left(\frac{2b}{a} \right) \text{ H m}^{-1}$$

In the discussion of inductance we have been concerned with the linkage of flux of a circuit with the current distribution producing the flux. Not surprisingly this is *self-inductance*. When flux associated with the current in one circuit links another circuit carrying an independent current then this is *mutual inductance*. Suppose that there are two independent coils with N_1 and N_2 turns carrying currents I_1 and I_2 respectively. If ψ_{12} is the flux linking one turn of the coil carrying I_1 due to I_2 then

$$L_{12} = \frac{N_1 \psi_{12}}{I_2}$$

and L_{12} is a mutual inductance. Similarly if ψ_{21} is the flux linking one turn of the coil carrying I_2 due to I_1 then

$$L_{21} = \frac{N_2 \psi_{21}}{I_2}$$

Sometimes the symbol M is used for mutual inductance.

Problems

4.1   A rectangular coil, carrying a current of 400π amperes, lies in the $z = 0$ plane with its corners at the points $(a, b, 0)$, $(-a, b, 0)$, $(-a, -b, 0)$ and $(a, -b, 0)$.
(a) Find an expression for the magnetic field strength H at a point $(x, y, 0)$ within the coil and evaluate it at points $(0, 0, 0)$ and $(2, 1\frac{1}{2}, 0)$ given $a = 4$, $b = 3$ metres.
(b) Find an expression for H at any point on the z axis and for the same coil evaluate it at $(0, 0, 2)$.

((a) 166·7, 234·7;   (b) 113·2 A/m)

4.2   A two-dimensional static vector field is given by

$$F = (x^2 - y^2)a_x + 2xya_y + 0a_z$$

Verify Stokes' theorem for the square defined by the points $(0, 0, 0)$, $(a, 0, 0)$, $(a, a, 0)$, $(0, a, 0)$. Could F be a field vector representing an electric or magnetic field strength?

$(2a^3$, no)

4.3   The current density in a conducting medium is $J = J_0 \exp(-|\rho|/a)a_z$ using cylindrical coordinates.
(a) Find the total current crossing the $z = 0$ plane in a circle radius ρ centred on the z axis.
(b) Find the electric fields strength H at ρ.
(c) Verify that curl $H = J$.

((a) $2\pi J_0 a\{a - (a + \rho) \exp(-\rho/a)\}$;   (b) $(J_0 a/\rho)\{a - (a + \rho) \exp(-\rho/a)\}$)

4.4   A single-turn plane coil of area $\Delta S = \Delta S a_z$ is placed at the origin and carries a current I. By finding the solid angle $\Delta \Omega$ subtended at any point defined by the radius vector $r = xa_x + ya_y + za_z$, show that the magnetic field strength at the point is

$$H = (I \, \Delta S/4\pi r^4)\{3za_r - ra_a\} \text{ A/m}$$

4.5   Two long parallel cylinders, radii a and spaced $2b$ apart, carry currents $\pm Ia_z$ amperes.
(a) Taking $A = 0$ between the conductors, and $b = 5a$, plot the curve of $|A|$ for points in the plane of the conductors, excluding points occupied by the conductor.

(b) What is the maximum value of $|A|$ and where does it occur?

(c) Use Stokes' theorem to find the flux through a square coil of sides $2c$, in the plane of and centrally between the conductors.

((b) $(\mu_0 I/2\pi) \ln (\rho_2/\rho_1) a_z, 2\cdot 4(\mu_0 I/2\pi)$ Wb/m at outer side of conductor; (c) $2c(\mu_0 I/\pi) \ln [(b + c)/(b - c)]$)

4.6   A parallel-wire transmission line carries currents $\pm I$; the distance between the wires is a. Derive for any point (a) the magnetic vector potential, (b) the allowed components of the magnetic field strength.

4.7   A circular current loop of radius a, carrying a current I, is centred on the origin such that its plane is in the xy plane. Calculate the magnetic vector potential A at an off-axis point which is a distance r from the centre of the loop. Show that this current loop produces a magnetic field whose variation is the same as the electric field from an electric field dipole at the origin.

$(H = (Ia^2/4r^3)(2 \cos \theta a_r + \sin \theta a_\theta))$

4.8   An infinitely long wire lies along the z axis and carries a current I. Derive an expression for the magnetic flux density at a distance r from the wire from the magnetic vector potential.

$(B = \mu_0 I a_\phi/2\pi r)$

4.9   Two coaxial solenoids have the following parameters: 100 turns per centimetre of 2 cm radius; 160 turns per centimetre of 3 cm radius. Calculate the mutual inductance of the pair.

4.10 An infinitely long coaxial line consists of an inner conductor of radius a and an outer conductor of inner radius b and thickness t. Currents $+I$ and $-I$ flow in the two conductors. Calculate the magnetic flux density at the following positions: (a) within the inner conductor, (b) between the two conductors, (c) inside the outer conductor.

$((c) (\mu_0 I/2\pi r) - (\mu_0 I(r^2 - b^2)/2\pi r[(b + t)^2 - b^2])$ for $b < r < b + t)$

5

Magnetic Forces and Magnetic Media

The basis of electrical machines is the interaction of the magnetic field and current-carrying conductor in producing mechanical force. In this context magnetic materials and magnetic circuits play an important part. Our previous discussion has been limited to magnetic effects, mainly in free space, which arise from conduction-type currents. Now, after first considering the mechanical forces exerted upon current-carrying conductors, we shall examine the basic magnetic behaviour of materials.

5.1 Forces due to magnetic fields

This is the second part of the problem stated by Ampère's equation given earlier. We have now dealt with the computation of magnetic flux density B (and field strength H) from the geometry of current-carrying conductors. It now remains to compute the mechanical forces and torques on current-carrying conductors placed in a magnetic field.

By comparison with the electric field strength E we have the following forces exerted by the magnetic field of flux density B.

	Force in electric field	Force in magnetic field
Stationary electron	$-q_e E$	zero
Moving electron	$-q_e E$	$-q_e u \times B$

where u is the velocity of the electron. For an electron in a region in which both electric and magnetic fields are present

$$\text{force } F = -q_e(E + u \times B)$$

$$= m_e \frac{\mathrm{d}}{\mathrm{d}t}(u)$$

where m_e is the mass of the electron. This is the Lorentz equation. The charge on the electron is negative and this determines the direction of the force. The free electron 'accelerates' along the direction of the electric field strength E; but the magnetic force is orthogonal to the motion and does not cause an increase in velocity but only a direction change. There is therefore no change in the kinetic energy of an electron due to the magnetic field, thus any energy transfer must be by means of the electric field.

5.1.1 Differential current element

The following expressions hold for the force experienced in a static magnetic field of flux density B.

$$dF = dq\,(u \times B) \qquad \text{moving charge } dq$$
$$= I\,dl \times B \qquad \text{filament element } I\,dl$$
$$= K\,db\,\,dl \times B = (K \times B)\,dS \qquad \text{sheet element } K\,dS$$
$$= J\,d(\text{vol}) \times B = (J \times B)\,d(\text{vol}) \qquad \text{volume element } J\,d(\text{vol})$$

On the atomic scale the application of a magnetic field to a current-carrying conductor (or semiconductor) will alter the electron motion and so too would a thermal gradient. A full discussion of these *galvanomagnetic effects* arising out of the interaction of particles of the system with electric and magnetic fields and thermal gradients would be out of place in this text (Putley, 1960). However, the appearance of an electric field in the y direction due to a current in the x direction, and a magnetic field in the z direction, the celebrated *Hall effect*, is well known. Also, owing to the fact that a magnetic field tends to curve the path of an electron, the conductivity of a material depends upon the magnetic field. This is known as *magnetoresistance*. In this book we shall be concerned with the forces listed above as acting on the conductor material.

5.1.2 Forces on a closed current-loop

A closed current-loop in a uniform magnetic field of flux density B is shown in figure 5.1a with the forces on each side of the loop indicated F_1, etc. We have for any element, $dF = I\,dl \times B$. Using this round the complete circuit

$$F = -I \oint B \times dl$$

This is the best we can do until we know the magnetic flux density B and the configuration. However, for a completely uniform flux density

$$F = -IB \times \oint dl$$

This loop integral $\oint dl$ must be 0 for any closed circuit, hence $F = 0$. This is typified in figure 5.1a for $F_1 = -F_3$, $F_2 = -F_4$ so that $F_1 + F_2 + F_3 + F_4 = 0$

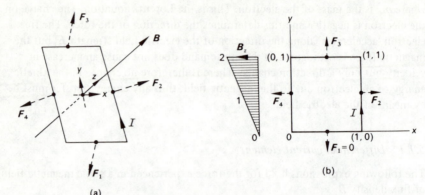

(a)

(b)

Figure 5.1   *Forces on current conductors in magnetic fields.*

when B is uniform. As an example of a coil in a non-uniform field, figure 5.1b, suppose $B = (2y)a_z$

$$F_1 = (I)1a_x \times (B = 0) = 0$$

$$F_2 = -F_4 = I(a_y) \times \tfrac{1}{2}(0 + 2)a_z = I(1a_x)$$

$$F_3 = I(-a_x) \times (2a_z) = 2I(+a_y) \neq F_1$$

Thus there is a y-directed force of translation $2I$ newtons.

5.1.3   *Torque on a current-carrying coil*

In a uniform field the vector sum of the forces in equilibrium is zero. However, there can still be a torque since the forces F_1, F_3 and F_2, F_4 in figure 5.1 form couples about the polar axis. We can compute the torque. Suppose an *elemental coil* dx, dy carries I and is placed in a field B, uniform over the coil: $B = B_x a_x + B_y a_y + B_z a_z$, refer to figure 5.2a where the plane of the coil is in the xy plane.

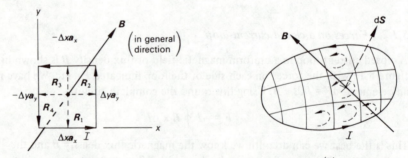

(a)

(b)

Figure 5.2   *Torque on current loop in magnetic field.*

The four forces are

$$\Delta F_1 = I\Delta x a_x \times B = I\Delta x(0a_x - B_z a_y + B_y a_z) = -\Delta F_3$$

$$\Delta F_2 = I\Delta y a_y \times B = I\Delta y(B_z a_x + 0a_y - B_x a_z) = -\Delta F_4$$

The torque arms are $R_1 = -\frac{1}{2}\Delta y a_y$; $R_2 = +\frac{1}{2}\Delta x a_x$, etc.

$$\Delta T = 2R_1 \times \Delta F_1 + 2R_2 \times \Delta F_2 = I\Delta x \Delta y(-B_y a_x + B_x a_y)$$

Noting that $a_z \times B = -B_y a_x + B_x a_y$ and using $(\Delta x \Delta y)a_z = \Delta S$ we can write generally

$$\Delta T = I\Delta S \times B$$

For a coil of any size and shape in any steady magnetic field of flux density B suppose the coil to be broken down into elemental meshes by filaments all carrying circulating currents I as we have illustrated in figure 5.2b. For any element, $dT = I\,dS \times B$. All these elemental couples will add vectorially and must by superposition produce the same torque as the current I round the periphery of S.

Hence $T = I\int_S dS \times B$ N m about the polar axis of the coil where presumably the coil may pivot.

If the field *is* uniform $T = -IB \times \int_S dS = -IB \times S$. This is the usual relation $T = -IBS \sin \theta$. The torque is usually so directed as to reduce θ so that S aligns with B, embracing the field flux.

5.2. Forces in circuits due to own fields

So far we have considered only mechanical forces produced on current-carrying conductors in magnetic fields created externally.

There are many cases where the mechanical force on a current-carrying conductor is due to the field created by the current in the rest of the circuit. A few typical cases follow.

5.2.1 Force between two parallel conductors

This is an important case, particularly for the single-phase transmission line. The two infinite conductors are shown in figure 5.3 and the element of length $dl = dza_z$.

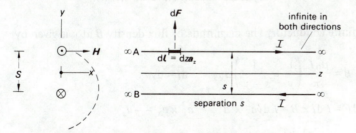

Figure 5.3 *Force between two parallel conductors.*

The magnetic field strength H, at A due to current I in B, is given by

$$H = \frac{I}{2\pi s}a_x; \qquad B = \frac{\mu_0 I}{2\pi s}a_x$$

The force dF on the current element $I\,dl$ is given by

$$dF = I\,dl \times B = I\,dza_z \times \frac{\mu_0 I}{2\pi s}\,a_x$$

$$= \frac{\mu_0 I^2}{2\pi s}\,dza_y$$

Hence

$$\frac{dF}{dx} = \frac{\mu_0 I^2}{2\pi s}\,a_y = \frac{2}{10^7}\,\frac{I^2}{s}\,a_y\ \text{N/m}$$

This is the Biot–Savart result which is used as the SI basis of the definition of the ampere.

5.2.2 Force on the blade of a circuit breaker

The description in figure 5.4 is of a circuit breaker; it is a simplification in which the blade CD is a sliding fit between conductors; this avoids the difficult computation of the effects at the corners. The field at y has a component from each

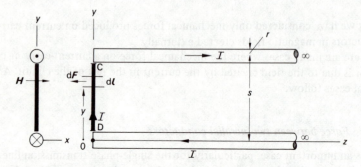

Figure 5.4 *Sliding bit circuit breaker.*

semi-infinite conductor. The magnitude of flux density B at y is given by

$$B = \frac{\mu_0 I}{4\pi}\left(\frac{1}{y} + \frac{1}{s-y}\right)a_x; \qquad dl = dya_y$$

$$dF = I\,dl \times B = I\,dya_y \times B; \qquad a_y \times a_x = -a_z$$

$$= \frac{\mu_0 I^2}{4\pi}\left(\frac{1}{y} + \frac{1}{s-y}\right)dy(-a_z)$$

$$F = \frac{\mu_0 I^2}{4\pi}\int_r^{s-r}\left(\frac{1}{y} + \frac{1}{s-y}\right)dy(-a_z) = \left[\frac{\mu_0 I^2}{2\pi}\ln\left(\frac{s-r}{r}\right)\right](-a_z)\ \text{N}$$

5.2.3   Hoop force in a flexible circular conductor

The flexible circular conductor is shown in figure 5.5 with an elementary current element $I \, \mathrm{d}l$ at radius R. The radius of the conductor is a and the current is I. Such a conductor experiences a radial force which tends to make it expand to embrace more self-flux. This produces a tension in the conductor akin to hoop stress. If we

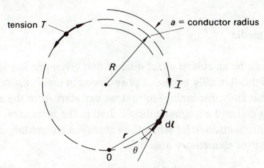

tension T — / a = conductor radius

Figure 5.5   *Hoop tension in a flexible conductor.*

try to compute this force T by finding the magnetic field strength H at the periphery due to the complete circuit by using the Biot–Savart rule

$$\mathrm{d}H = \frac{I \, \mathrm{d}l}{4\pi r^2} \times a_r$$

difficulties arise as $\mathrm{d}l$ approaches the point (say O) for, though $\mathrm{d}l \times a_r \to 0$ as $\theta \to 0$, $1/r^2 \to \infty$ at a greater rate and the loop integral cannot be simply found.

Another method is to consider the change ΔW in stored energy as the radius R increases by ΔR and I is constant

$$W = \tfrac{1}{2}LI^2 \qquad \Delta W = \tfrac{1}{2}I^2 \Delta L = T 2\pi \Delta R$$

Hence

$$T = \frac{I^2}{4\pi} \frac{\Delta L}{\Delta R}$$

For such a circular conductor we can use Rayleigh's formula (Jeans, 1925) for the self-inductance of a conductor in a circular loop

$$L = \mu_0 R \left[\left(1 + \frac{a^2}{8R^2} \right) \ln \left(\frac{8R}{a} \right) + \frac{a^2}{24R^2} - \frac{7}{4} \right]$$

at low frequency

$$\approx \mu_0 R \left[\ln \left(\frac{8R}{a} \right) - \frac{7}{4} \right] \qquad \text{when } R \gg a$$

$$\frac{\Delta L}{\Delta R} = \mu_0 \left[\ln\left(\frac{8R}{a}\right) - \frac{7}{4} + R \frac{8}{a} \frac{a}{8R} \right] = \mu_0 \left[\ln\left(\frac{8R}{a}\right) - \frac{3}{4} \right]$$

Hence

$$\text{hoop tension } T = \frac{I^2}{4\pi} \mu_0 \left[\ln\left(\frac{8R}{a}\right) - \frac{3}{4} \right] \text{ N}$$

5.3 Magnetic media

The treatment can be an almost exact dual of that given previously for dielectrics in the electric field. Basically magnetic phenomena in media are of current origin (electrons in orbit and/or spinning) so that we can start from the equivalence of a current-carrying coil and a magnetic dipole shell in the same area. This has been discussed earlier in section 4.7.1 when the equivalent elementary magnetic moment dm was obtained for elementary area dS_m.

$$dm = \mu_0 I \, dS_m$$

The equivalence of bound current and magnetic shell is illustrated in figure 5.6 which consists of n shells and n current loops.

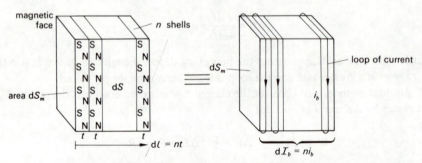

Figure 5.6   *Equivalence of magnetic shell and bound current.*

Referring to figure 5.6 for a small bound current i_b flowing round an elementary area $d(S_m)$, of thickness t, the equivalent magnetic moment $dm = \mu_0 i_b \, d(S_m)$.

Now we define intensity of magnetisation M in a medium as

$$M \equiv \frac{\text{total magnetic moment}}{\text{volume}}$$

Consider the elementary volume of n shells, shown above and its equivalent bound current $d(I_b) = n i_b$. Its total magnetic moment is

$$M \, d(\text{vol}) = n\mu_0 i_b \, d(S_m) = \mu_0 \, d(I_b) \, d(S_m)$$

Suppose next we are considering $\oint H \cdot dl$ in the magnetised medium shown in figure 5.7a in which the elementary volume lies within dl and dS_m, that is $d(vol) = dS_m \cdot dl$.

After magnetisation there is current alignment as indicated in figure 5.7b. We can write $M(dS_m \cdot dl) = \mu_0 \, dI_b \, dS_m$. From this $M \cdot dl = \mu_0 \, dI_b$. To obtain the total bound current round S_m or through S

$$I_b = \frac{1}{\mu_0} \oint_S M \cdot dl$$

This integral is to be taken round the complete magnetic field strength loop which we are considering; it is around S, not S_m. If J_b represents the bound current density in the material, then

$$\int_S J_b \cdot dS = I_b$$

Note that S here refers to the magnetic field strength contour and should not be confused with the magnetised surface S_m above, which is no longer needed.

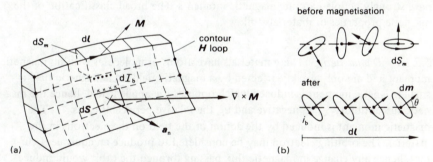

Figure 5.7   *Bound current in magnetic material.* (a) $\oint H \cdot dl$ *in magnetised medium.* (b) *Current alignment upon magnetisation.*

From this, and using Stokes' theorem we can write

$$\frac{1}{\mu_0} \oint_S M \cdot dl = I_b = \int_S J_b \cdot dS = \frac{1}{\mu_0} \int_S (\nabla \times M) \cdot dS$$

Hence the relation $(1/\mu_0)(\nabla \times M) = J_b$.

Now in free space (before the advent of the material) we have $\nabla \times H = (J) = J_f$. This can be put in terms of B as $\nabla \times (B/\mu_0) = J_f$. But, when a material is introduced as a medium, B changes so that $\nabla \times (B/\mu_0) = J_f + J_b = J = \nabla \times H + (1/\mu_0) \nabla \times M$. Finally $B = \mu_0 H + M$.

But M is linearly related to H (its actual cause) so that following the electric-field analysis we write $M = \chi_m \mu_0 H$, where χ_m is the magnetic susceptibility, which produces

$$B = \mu_0(1 + \chi_m)H = \mu_0 \mu_r H = \mu H$$

The term $\mu_r = 1 + \chi_m$ is the *relative permeability* of the material.

5.3.1 The atomic theory of magnetism

A full treatment of the atomic and quantum theory of magnetism is beyond the scope of this book but a brief excursion into elementary solid state physics will be given. The simple model of an atom is of electrons orbiting a nucleus while spinning on their own axis. Each of these effects contributes to the magnetic moment of the atom, giving an orbital magnetic moment and a spin magnetic moment respectively. A further magnetic effect due to nuclear spin may safely be neglected at this level. The overall magnetic moment of an atom is due to the resultant of the spin and orbit moment and even for a 'free' atom this is usually complicated. When the atoms come together in the solid state, the atomic magnetic moments are modified and some materials have atoms with zero resultant magnetic moment, some have little coupling between the magnetic moments on individual atoms, whilst others have strongly coupled atomic magnetic moments. The broad classification of the magnetic properties of materials follows.

5.3.1.1 Diamagnetic

These materials have atoms or molecules with no magnetic moment and are only weakly affected by a magnetic field. In a magnetic field a small magnetic moment is induced which opposes the applied field, hence the magnetic susceptibility χ_m is negative and μ_r, the relative permeability, is <1. The magnetic moment is induced by the action of the field on the electrons in the material. The orbiting electron may be considered to produce an equivalent current, hence any change in magnetic flux passing through the orbit would induce another equivalent current in the orbit which by Lenz's law (see chapter 6) opposes the change in flux. The induced current corresponds to a change in angular velocity of the orbiting electron and the induced magnetic moment opposes the applied field. Examples of diamagnetic materials include copper, gold, hydrogen gas, etc. Of course *all* materials exhibit diamagnetism but in many materials this is swamped by other magnetic effects.

5.3.1.2 Paramagnetic

These are materials in which the atoms have a resultant magnetic moment. In an applied field these elementary magnets rotate towards the direction of the applied field and the resultant magnetic moment depends upon the ability these have to line up with the applied field. This lining up is dependent upon the thermal energy within the material which tends to randomise the orientation. Indeed the calculation of the overall magnetic moment is a relatively simple

exercise in quantum statistics. Paramagnetic substances have a positive susceptibility and a value of relative permeability just >1. The susceptibility is temperature-
dependent and is far greater than the diamagnetic contribution. The main applications of paramagnetic materials have been as the active material in masers
(*microwave amplification by stimulated emission of radiation*).

5.3.1.3 Ferromagnetic In these materials the relative permeability is very large,
$\mu_r \gg 1$, and is field-dependent. Below a certain critical temperature, the *Curie
temperature*, a ferromagnetic material can exhibit spontaneous magnetisation, that
is magnetisation in zero applied field. Ferromagnetic materials exhibit hysteresis—
changes in magnetic flux density lagging behind changes in applied field—which
may or may not be of use to the designer! The magnetic moments of individual
atoms are strongly coupled to neighbouring atoms by the *Weiss molecular field*, a
strong internal field. This field arises from a quantum-mechanical interaction
between electrons in neighbouring atoms, giving strong coupling between electron
spins which is termed *exchange interaction.*

As a result of this, small volumes in a material are magnetised in a particular
direction. The volumes are called domains and, in each, the magnetisation is satu-
rated. However, the domains might, in zero field, be completely randomly orien-
tated, giving zero magnetisation on a macroscopic scale. When such a material in its
unmagnetised state is magnetised the domain structure is changed. Those in the
direction of the field grow at the expense of less favoured domains. At first the
growth is reversible, then irreversible, finally domain orientation takes place and the
material reaches saturation. This latter process is difficult, and reducing the applied
field to zero has only a small effect on the magnetic flux density, which can be
reduced to zero only by reversing the field. Thus it is easy to explain the hysteresis
cycle by domain growth and movement.

Ferromagnetism occurs in iron, nickel and cobalt at room temperature and in
various alloys, some of which do not contain a magnetic element! It is not necessary
to expand on the usefulness of ferromagnetism in electrical engineering.

5.3.1.4 Antiferromagnetism Although strong interaction occurs the magnetic
moments of neighbouring atoms assume an anti-parallel arrangement. Hence a
weakly magnetic material is typified by manganese oxide.

5.3.1.5 Ferrimagnetism In complicated structures, such as spinels, there can be
more than one magnetic atom site. The ferrites have this, and other structures,
with atoms of differing moments on the sites in an anti-parallel manner. The result
is a magnetic material exhibiting hysteresis effects but, because of their chemical
composition (such as $Mn_x Zn_{1-x} Fe_2 O_4$) based on metallic iron oxides, having useful electrical characteristics. These include low conductivity and hence the use of
ferrites at high frequencies.

For further discussion on magnetic materials the reader is advised to refer to one
of the specialist texts (Morrish, 1965; Chikazumi, 1965).

5.4 Boundary conditions for magnetic media

This follows closely the dielectric case previously considered, with one notable exception due to the likelihood of surface current at the boundary, such as air gap conditions in machines. Assume the two media to be of isotropic and homogeneous materials. Next, suppose the surface current density K A/m to be orthogonal to the

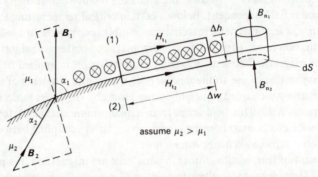

Figure 5.8 *Boundary conditions between two magnetic media.*

surface tangent, that is conductors *not* skewed. This is illustrated in figure 5.8; the suffixes n and t refer to normal and tangential components respectively. For the tangential conditions consider an Ampère loop of width Δw and height Δh negligible; proceeding clockwise

$$-H_{t_2}\Delta w + H_{t_1}\Delta w = K\Delta w$$

from which

$$H_{t_1} - H_{t_2} = K \text{ A/m}$$

Next, for the normal conditions, consider the gaussian pillbox, and use Maxwell's fourth equation $\oint B . dS = 0$. We can again make Δh negligible. For the ends, area dS

$$(dS)B_{n_1} - (dS)B_{n_2} = 0$$

or

$$B_{n_1} = B_{n_2}$$

Suppose $\mu_1, \mu_2, B_2, H_2, \alpha_2, K$ are all known. To obtain B_1 use the relations just found and finally

$$B_1 \sin \alpha_1 = B_2 \sin \alpha_2$$

$$\frac{B_1 \cos \alpha_1}{\mu_1} = K + \frac{B_2 \cos \alpha_2}{\mu_2}$$

$$\cot \alpha_1 = \frac{\mu_1}{\mu_2}\left(1 + \frac{K\mu_2}{B_2 \cos \alpha_2}\right)\cot \alpha_2$$

Of course with ferromagnetic materials, not only is the material anisotropic (permeability different in different directions) but μ_r is not a constant and depends on the flux density B.

5.5 Energy storage in the magnetic field

This is not so easily deduced as the electric counterpart, because now the fields are not conservative and changes in them may cause changes in the current source. The easiest way is to assume the expression for the energy stored in an inductive circuit

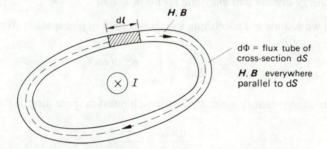

Figure 5.9 *Energy stored in magnetic field.*

$W_M = \frac{1}{2}LI^2$. For a one-turn circuit as described by figure 5.9 this becomes

$$W_M = \tfrac{1}{2}(\lambda I) = \tfrac{1}{2}(\Phi I)$$

Suppose we consider a constant-current source; for flux changes there will be voltage changes to keep the current constant. For this

$$dW_M = \tfrac{1}{2}(d\Phi)I = \tfrac{1}{2}(B \cdot dS)I = \tfrac{1}{2}(B \cdot dS) \oint H \cdot dl$$

If we consider an elementary flux tube as shown, in taking the loop integral we must proceed round one complete tube in which B, H and dl are each parallel to dS. For this tube

$$dW_M = \tfrac{1}{2} \oint BH \, d(vol)$$

taken round tube. For all the tubes, that is all space, this becomes

$$W_M = \tfrac{1}{2} \int_{vol} B \cdot H \, d(vol)$$

This is perfectly comparable with

$$W_E = \tfrac{1}{2} \int_{vol} D \cdot E \, d(vol)$$

for the electric field and so we can say for the general magnetic field where there are no restrictions on B and H as space vectors

$$\text{stored magnetic energy } W_M = \tfrac{1}{2} \int_{\text{vol}} (B.H) \, \mathrm{d(vol)} \; \text{J}$$

In anisotropic media, B and H are not necessarily directed alike and the magnetic susceptibility has to be described by a tensor.

5.5.1 Energy density and magnetic vector potential

The total work done in establishing a magnetic field in free space is given by

$$W_M = \frac{1}{2\mu_0} \int_{\text{all space}} B^2 \, \mathrm{d(vol)}$$

so that an energy density associated with each point in space may be defined as

$$\frac{\mathrm{d}W_M}{\mathrm{d(vol)}} = \frac{B^2}{2\mu_0}$$

Referring again to the expression for the magnetic flux density B in terms of magnetic vector potential A

$$B = \nabla \times A$$

thus the expression for stored energy may be written

$$W_M = \frac{1}{2\mu_0} \int_{\text{vol}} (B.\nabla \times A) \, \mathrm{d(vol)}$$

Now

$$B.\nabla \times A = \nabla.(A \times B) + A.(\nabla \times B)$$

and from this and the divergence theorem

$$W_M = \frac{1}{2\mu_0} \int_{\text{vol}} A.(\nabla \times B) \, \mathrm{d(vol)} - \frac{1}{2\mu_0} \int_{\text{vol}} A \times B.\mathrm{d}S$$

The surface integral vanishes if the volume includes all space and the surface S is at infinity. Thus, if $\nabla \times B = \mu_0 J$, where J is the current density

$$W_M = \tfrac{1}{2} \int_{\text{vol}} (J.A) \, \mathrm{d(vol)}$$

where the volume includes all space for which J is non-zero.

5.6 Mechanical force on magnetic material

Here one is concerned with the mechanical force acting on a magnetic material, for this has obvious practical importance in the design of electrical machines, electro-magnets, etc.

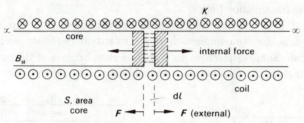

Figure 5.10 *Mechanical force on electromagnetic core faces.*

One usual treatment is to consider the infinite solenoid carrying K A/m $= H_K$, as is shown in figure 5.10. In this the cores of the solenoid are assumed to be separable as we have indicated. The flux density in the steel cores is

$$B_{st} = \mu H_K = \mu_r \mu_0 H_K$$

Move the cores apart under the action of forces F between faces thus creating an air gap, in which the flux density is B_a and the field intensity H_a, then

$$F\,dl = \text{energy stored in gap} = \tfrac{1}{2} B_a H_a (dl)(S)$$

$$F/S = \tfrac{1}{2} B_a H_a \text{ newtons per square metre of cross-sectional area}$$

The difficulty is in knowing what are the flux density B_a and field strength H_a in the gap. If we assume $B_a = B_{st}$ in the air gap then $H_a = B_{st}/\mu_0$

$$F/S = \tfrac{1}{2} B_{st}^2/\mu_0 \text{ N/m}^2 = \tfrac{1}{2}\mu_0 H_a^2 \text{ N/m}^2$$

This is patently *wrong* for if the core were wooden ($\mu_r = 1$) there would still apparently be a magnetic force on it!

The difficulty can be overcome by noting that after creating the gap, magnetised surfaces ($M = \chi_m/\mu_0 H$) have appeared, giving rise *within* the material to an internal countering force $\tfrac{1}{2} B_{st}^2/\mu$ N/m^2.

Corrected net force on one face $= \tfrac{1}{2} B_{st}^2 \left(\dfrac{1}{\mu_0} - \dfrac{1}{\mu} \right)$ N/m

$$F/S = \frac{1}{2} \frac{B_{st}^2}{\mu_0} \left(1 - \frac{1}{\mu_r} \right) = \frac{1}{2} \frac{B_{st}^2}{\mu_0} \frac{\mu_r - 1}{\mu_r}$$

$$= \tfrac{1}{2} H_K^2 \mu_0 \mu_r (\mu_r - 1)$$

These expressions clearly vanish for a non-magnetic core ($\mu_r = 1$.) For iron where μ_r ranges from 100 to 100 000, the correction is unimportant.

5.7 Magnetic circuit

This is treated far less than the current circuit, for a magnetic circuit is mostly of concern to the designer. Magnetic circuits occur in fixed items such as transformers; they are not distributed countrywide or readily alterable as are the current networks such as transmission lines.

They are all based on Ampère's circuital law with some assumptions regarding flux flow and, as the name implies, there is similarity between this and the d.c. resistive circuit. Some differences do arise, however, because of the non-linear magnetic effects in the ferromagnetics which form part of the circuit.

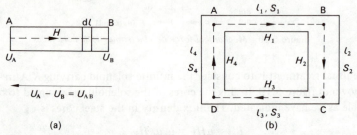

Figure 5.11   *Magnetic circuit (idealised).* (a) *One limb.* (b) *Complete circuit.*

Consider one limb of a magnetic circuit, see figure 5.11a, the complete circuit being shown in figure 5.11b. The magnetic scalar potential difference between A and B is

$$U_B - U_A = - \int_A^B H . \, \mathrm{d}l$$

$$U_{AB} = U_A - U_B = \int_A^B H . \, \mathrm{d}l$$

Hence for a complete circuit $U_{AB} + U_{BC} + \ldots = \oint H . \, \mathrm{d}l = IN = U_{mmf}$ in which N is the number of turns. U_{mmf} is called the magnetomotive force (or m.m.f.) a nomenclature which is misleading for like electromotive force (e.m.f.) *it is not a force.* The units of magnetomotive force are amperes, or ampere turns for a coil with many turns. For the complete circuit of figure 5.11b

$$U_{mmf} = IN = U_{AB} + U_{BC} + U_{CD} + U_{DA} = H_1 l_1 + H_2 l_2 + H_3 l_3 + H_4 l_4$$

$$= \frac{B_1 l_1}{\mu_1} + \frac{B_2 l_2}{\mu_2} + \frac{B_3 l_3}{\mu_3} + \frac{B_4 l_4}{\mu_4} = \frac{\Phi_1 l_1}{\mu_1 S_1} + \frac{\Phi_2 l_2}{\mu_2 S_2} + \frac{\Phi_3 l_3}{\mu_3 S_3} + \frac{\Phi_4 l_4}{\mu_4 S_4}$$

$$= \Phi \left(\frac{l_1}{\mu_1 S_1} + \frac{l_2}{\mu_2 S_2} + \frac{l_3}{\mu_3 S_3} + \frac{l_4}{\mu_4 S_4} \right) = \Phi (R_{m1} + R_{m2} + R_{m3} + R_{m4})$$

This assumes there is no flux leakage, so $\Phi_1 = \Phi_2 = \ldots$.

The ratio of the magnetomotive force to the total flux is termed the magnetic resistance or reluctance R_m and $R_m = U_{mmf}/\Phi$. Hence if we knew the flux Φ required, and the permeabilities of the parts, we could compute the reluctances and then the magnetomotive force. Theoretically the m.m.f. should be distributed round the circuit but for engineering convenience it is mostly put as a coil, carrying current, on one limb.

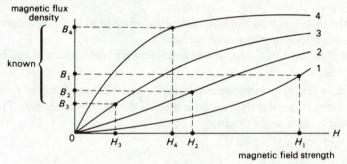

Figure 5.12 *Composite magnetic circuit with known flux.*

This treatment is rarely of use to anyone, least of all a designer, for ferromagnetic materials are notoriously non-linear; μ values are not determinable until either the flux density or magnetic field strength is known. As an example of the design principle used let us assume the simple magnetic circuit of 5.11b in which the lengths l and areas S are known, as too is the flux Φ. Figure 5.12 shows the variation of magnetic flux density B with magnetic field strength H for the four limbs, and the calculations are tabulated below (table 5.1).

Table 5.1

B	Φ/S_1	Φ/S_2	Φ/S_3	Φ/S_4	Computed from flux
H	H_1	H_2	H_3	H_4	Obtained from curves
Hl	H_1l_1	H_2l_2	H_3l_3	H_4l_4	Computed from lengths
m.m.f. $= IN = U_1 +$	$U_2 +$	$U_3 +$	U_4		Total

Even when so computed, there may be 10 to 15 per cent errors due to several factors

(i) The lengths l_1, l_2, l_3, l_4 (figure 5.11b) are never so distinct, since the effect of the 'corners' is important.

(ii) The core is made up by overlapping or butting flat plates, and thus there will be an inevitable air gap.

(iii) There will be a fair leakage of flux across the air from points of high magnetic potential U to points of low potential.

(iv) The effect of concentrating m.m.f. in one limb could be too gross an assumption.

Some of these sources of error can be allowed for by introducing empirical factors in the form of percentage increases in IN for leakage, corners, joints.

The three-limbed three-phase core brings additional difficulties, owing to the asymmetry (middle limb disposed differently from outers). Several assumptions are also made regarding the magnetic properties of the materials; these include

(i) No variation in relative permeability occurs in each type of material in the circuit.

(ii) The magnetic field and flux density are assumed to be either tangential or normal to all ferromagnetic surfaces.

(iii) It is assumed that through any cross-section of the material B and H are constant.

It is worth ending by pointing out that the reluctance of any air gap in a magnetic circuit is often far greater than the reluctance of the ferromagnetic part. This is of obvious importance to the designer.

Problems

5.1   A rigid rectangular circuit in the $z = 0$ plane has corners at $(0, 0, 0)$ $(1, 0, 0)$ $(1, 2, 0)$ $(0, 2, 0)$ and carries a current of 50 A anti-clockwise. Find the total force on the loop and the torque about its polar axis when it is in an extensive, magnetic field of strength H given by (a) $(5a_x + 2a_y - 4a_z)/10$, (b) $(2xa_x - 2ya_y)/10$ and (c) $(2ya_x + 2a_y + 2xa_z)/10$ tesla.

((a) 0, $10(-2a_x + 5a_y)$ N m, $10(2a_x + 1a_y)$ N m, $10(-2a_x + 2a_y)$ N m)

5.2   A rigid circular coil enclosing area 0.1 m^2 carries a current of 100 A and lies in a plane defined by $8x - y + 4z = 9$. A uniform extensive magnetic field of flux density B of $(9a_x + 6a_y - 6a_z)/10$ tesla surrounds the coil. (a) Compute the torque on the coil. (b) Show that if the coil were moved to another plane $6x + 4y + 13z = (221)^{1/2}$ the torque on it would be a maximum and find its value.

((a) $1/3(-6a_x + 28a_y + 19a_z)$, 11.5 N m,
(b) $(1/\sqrt{221})(-102a_x + 153a_y + 0a_z)$, 12.4 N m.)

5.3   Figure P5.3 shows the contact arm RS of a switch carrying 10 kA under

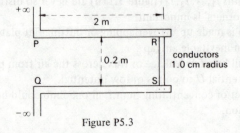

Figure P5.3

fault conditions. Find the force on the contact arm taking all conductors as cylinders of radius 1·0 cm. (Assume RS is a sliding fit in PR, QS.)

(59 newtons (approx.),   57.1 newtons (more accurate),

5.4   Two extensive superparamagnetic materials 1, 2 have a boundary surface at $x = 0$ plane. For $x < 0$, $\mu_1 = 1 \cdot 5\mu_0$ and for $x > 0$, $\mu_2 = 6\mu_0$. In the first region $B_1 = (10a_x + 4a_y - 3a_z)/20$ tesla. Find (a) the flux density B_2 and the ratio $|B_2/B_1|$; (b) the angle between B_1 and B_2; (c) the energy densities in J/m^3 in each material.

((a) $(10a_x + 16a_y - 12a_z)/20$, 2/1;   (b) $30°$;   (c) $5\mu_0/48$)

5.5   Two magnetic materials are separated by a planar interface which lies in the $x\,y$ plane. The material below the plane (material 1) has a permeability $\mu_{r1} = 5$ and the material above the $x\,y$ plane (material 2) has a permeability $\mu_{r2} = 15$. There is a current sheet $K = (13/100\mu_0)\,a_x$ A/m orthogonal to the tangential components of the magnetic field at the interface. If the flux density in material 1 is $B_1 = (0a_x + 16a_y + 12a_z)/20$
(a) Find the flux density B_2 on material 2.
(b) What must be the value of K to make B_2 entirely normal?

((a) $(0a_x + 9a_y + 12a_z)/20$ tesla;   (b) $(16/100\mu_0)$ A/m)

5.6   A very simple model of an atom may be considered to be a heavy nucleus of charge $+q_e$ around which an electron of charge $-q_e$ moves in a fixed Bohr orbit of radius a. If the electron has mass m_0 and a magnetic flux density of B is applied perpendicular to the orbit, show that there is a change in angular velocity ω_0 of $q_e B/2m$. Calculate the dipole moment in the absence of the applied magnetic flux density and the change in magnetic moment brought about by the application of the magnetic field.

$(q_e\omega_0 a^2/2, \quad q_e^2 a^2 B/4m_0)$

5.7   A uniform magnetic flux density B_2 exists in the z direction and a uniform p-type semiconductor of rectangular cross-section is placed in the magnetic field with its axis in the x direction. A current density J_x flow along the axis of the semiconductor which has charge carrier density p. Show that an electric field E_y exists in the y direction. This is the Hall effect and the Hall constant R_H is defined as the ratio $E_y/J_x B_z$. Show that for this case $R_H = 1/pq_e$.

5.8   Show that the Hall constant R_H of a specimen carrying both electrons and holes is given by

$$R_H = \frac{p\mu_p^2 - n\mu_n^2}{(p\mu_p + n\mu_n)^2 q_e}$$

where p and n are the hole and electron charge carrier densities and μ_p and μ_n the respective mobilities.

5.9   A coaxial cable has inner radius a and outer radius b. Calculate the self-inductance of the cable by considering the total magnetic energy stored when a current I is carried. Check the answer by repeating the calculation in terms of flux linkage.

$(L = (\mu_0/2\pi)[(\tfrac{1}{4}) + \ln(b/a)])$

5.10   An iron ring has a cross-sectional area of 10 cm^2 and a mean circumference of 50 cm. It is required to produce a magnetic flux density of 1 tesla and for this particular specimen a prior determination of the magnetic properties indicate that a magnetic field strength of 500 A/m is required to produce this flux density. Calculate (a) the reluctance of the ring and (b) the number of ampere turns required. Repeat these calculations for a similar ring with an air gap of width 5 mm. Find the force between the two faces of the air gap.

6

Fields Varying in Time

Previously our discussion has been concerned with steady or static charges and steady or direct currents. The fields were electrostatic and magnetostatic. Now we study the effects of time variations of charges, currents and potentials. Of particular interest will be fields arising from sinusoidal variations at both low and high frequency. The effect of varying fields, in time, on Maxwell's equations will be to leave the divergence equations unaltered but to modify the curl equations. This modification is due to (i) the electric field E produced by a changing magnetic field B which derives from Faraday's experimental law and (ii) an inspired hypothesis of Maxwell which modifies Ampère's circuital law to include the time variation of the electric flux density.

The time variation of the electric flux density is known as the displacement current density. It is not associated with the flow of a physical quantity and need not have a physical meaning associated with it. It is a property of the electromagnetic field.

6.1 Faraday's law

The importance of Faraday's experimental observations on induced currents cannot be overemphasised. These experiments, which were carried out in the 1830s, and independently by Henry in the United States at the same time, are the basis of electrical engineering. The results of the experiments may simply be stated that a magnetic field varying in time produces an electromotive force which causes a current to flow in a closed circuit in the field.

Since there is no current flow when the magnetic field is constant, the varying magnetic field has an effect on the electrons similar to the effect of a potential drop through a circuit. The time-varying field produces a voltage in the closed circuit which causes current flow. This voltage is called the electromotive force or e.m.f., but remember it *is not a force.*

The electromotive force V_{emf} which is induced in a closed circuit is proportional to the time rate of decrease of magnetic flux linkage in the circuit.

$$V_{\mathrm{emf}} = -\frac{\mathrm{d}\lambda}{\mathrm{d}t} = -N\frac{\mathrm{d}\Phi}{\mathrm{d}t} \qquad \text{(volts)} \quad \text{scalar}$$

The negative sign (due to Lenz) emphasises that the direction of the possible current flow is such as to tend to prevent the flux change taking place.

From now on we shall not continue with turns N, and simply concentrate on Φ assuming that only singly looped circuits are considered. Strictly we should use λ.

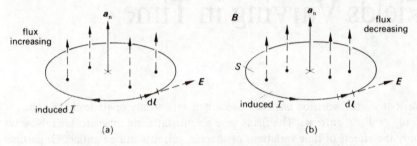

Figure 6.1   *Induced e.m.f.; stationary conductor, changing flux.*

The e.m.f. in the circuit V_{emf} is a *scalar*; it is of the same nature as potential difference and is measured in *volts*, as a dimensional check will reveal. A single loop circuit is shown in figure 6.1, for two cases: (a) for an increase in flux and (b) for a decrease. The figure also shows the assumed positive direction for the surface S and the unit normal vector a_{n}, the path element $\mathrm{d}l$ and the electric field strength E. The path must be closed so that

$$V_{\mathrm{emf}} = \oint_{l} E.\,\mathrm{d}l = -\frac{\mathrm{d}\Phi}{\mathrm{d}t} = -\frac{\mathrm{d}}{\mathrm{d}t}\int_{S} B.\,\mathrm{d}S \text{ volts}$$

In figure 6.1a the flux is increasing, V_{emf} is negative and the actual current I flows opposite to $\mathrm{d}l$ and E; in the converse case V_{emf} is positive and I flows in the direction of $\mathrm{d}l$ and E as shown in figure 6.1b.

From the equation it is seen that E now has *circulation* round the path (in the static case it had none). For more than one turn we could consider the surface S as re-entrant, thus including the turns as explained earlier.

The variation of flux linkage with time can be by numerous and devious ways. Two only are generally of practical importance.

(i)   *Stationary path*   The path shown in figure 6.1a and b is fixed and the flux linking it is varied by, say, current variations or magnet movements elsewhere. Since S is not itself changing with time we can take $\mathrm{d}/\mathrm{d}t$ inside the integral and associate it solely with the magnetic flux density B; then use Stokes' theorem

$$V_{\text{emf}} = \oint_l E . \, dl = - \int_S \frac{\partial B}{\partial t} . \, dS$$

$$\int_S \nabla \times E . \, dS = - \int_S \frac{\partial B}{\partial t} . \, dS$$

so that finally

$$\nabla \times E = - \frac{\partial B}{\partial t} \qquad \begin{array}{l} \text{completed Maxwell second} \\ \text{equation} \end{array}$$

In power engineering this V_{emf} is the 'transformer' e.m.f.

(ii) *Flux density constant; moving path*   A moving path is shown in figure 6.2 and the rails and bar are considered to be perfect conductors. Hence $\oint E . \, dl$ is zero along the circuit except at the voltmeter.

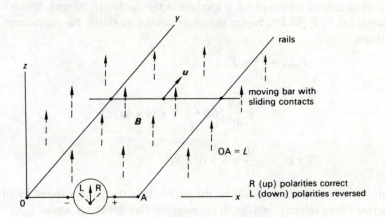

Figure 6.2   *Induced e.m.f.; moving conductor, constant flux density.*

At any instant $\Phi = BLy$ assuming the magnetic flux density B to be uniform and constant

$$V_{\text{emf}} = - \frac{d\Phi}{dt} = - BL \frac{dy}{dt} = - BLu$$

The moving bar will have charges in it, say Q coulombs. The mechanical force experienced by it is $F = Qu \times B$ and the electric field strength produced in it becomes

$$F/Q = E_{\text{m}} = u \times B \quad \text{('motional' electric field)}$$

and for the 'motional' e.m.f.

$$V_{\text{emf}} = \oint_l E_{\text{m}} \cdot dl^* = \oint_l (u \times B) \cdot dl^*$$

Note especially that, although the integral is shown round the closed loop, only those parts which are moving are involved; these are indicated by the asterisk dl^*. Also the sign of $u \times B$ gives the direction of the induced motional e.m.f. As shown in figure 6.2 the voltmeter will read 'upscale' (assuming the pointer to move to the right or 'upscale' when the applied potential difference agrees with the terminal polarity markings).

Combining the above results; for a moving conductor and changing magnetic flux

$$V_{\text{emf}} = \oint_l E \cdot dl = -\frac{\partial}{\partial t} \int_S B \cdot dS = -\frac{d\Phi}{dt}$$

$$= -\int_S \frac{\partial B}{\partial t} \cdot dS + \oint_l (u \times B) \cdot dl^*$$

If the whole of the periphery is involved in the 'motional' integral, Stokes' theorem can be applied to two of the terms, producing finally the point-form equation

$$V_{\text{emf}} = \int_S \nabla \times E \cdot dS$$

$$= -\int_S \frac{\partial B}{\partial t} \cdot dS + \int_S \nabla \times (u \times B) \cdot dS$$

$$\nabla \times E = -\frac{\partial B}{\partial t} + \nabla \times (u \times B)$$

In this latter case the field would be the electric field strength measured by an observer whose velocity, relative to the magnetic flux density B, was u.

(iii) *Classic problems* There are many 'classic' cases of difficulty in applying these relations, mostly arising from sliding contacts, switches and circuit substitution. Many workers since the time of Faraday have been concerned with theorising and experimenting about e.m.f. production including Hering, Cramp, Howe, Bewley, Cullwick and, most recently, Laithwaite. Cramp (1933–6) carried out some fifty experiments, all trying to decide whether or not the lines (tubes) of force of a bar magnet rotated when the magnet rotated; the results were inconclusive. Howe maintained it was a meaningless question and you could take your choice! Figure 6.3 illustrates five classic problems. Two only give electromotive force, and the reader should spend some time before deciding which two! In each case the magnet is assumed to have a cylindrical shape.

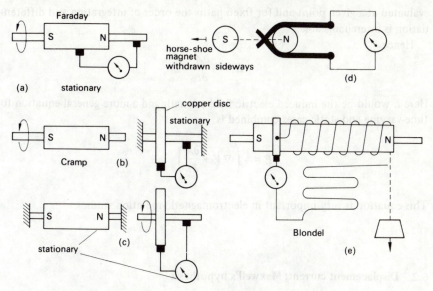

Figure 6.3   *A few classic problems on induced e.m.f.*

6.1.1   *Faraday's law and vector potential*

The magnetic flux Φ through a closed surface S is

$$\Phi = \int_S B \cdot dS$$

In terms of vector potential A this may be rewritten

$$\Phi = \int_S \nabla \times A \cdot dS$$

thus by the use of Stokes' theorem

$$= \oint_l A \cdot dl$$

According to Faraday's law, the *induced* electric field strength E is given by

$$\oint_l E \cdot dl = -\frac{d\Phi}{dt} = \frac{-d}{dt} \oint_l A \cdot dl$$

$$= -\oint_l \frac{\partial A}{\partial t} \cdot dl$$

The partial derivatives are used because the vector potential's time derivative is

evaluated at a given point and for fixed paths the order of integration and differentiation is interchangeable.

Hence

$$E = -\frac{\partial A}{\partial t}$$

Here E would be the induced electric field strength and a more general equation for
time-varying and static cases combined is

$$E = -\left(\nabla V + \frac{\partial A}{\partial t}\right)$$

This equation is very important in electromagnetic-radiation studies.

6.2   Displacement current; Maxwell's hypothesis

Starting from Faraday's experimental law we obtained a modified form of Maxwell's
equation, for curl of the electric field strength E

$$\text{(space)} \quad \nabla \times E = -\frac{\partial}{\partial t} B \quad \text{(time)}$$

This is a time-changing magnetic flux density B influencing or 'producing' a space-
changing electric field strength E. What of the converse? Should a time-changing
electric field produce a space-changing magnetic field? This was Maxwell's hypothesis which lead him to formulate the possibility of electromagnetic waves in space
some twenty-five years before they were physically produced by Hertz.

By analogy with the curl E equation we might expect the relation

$$\nabla \times H = -\frac{\partial}{\partial t} D$$

Let us see if this can be possible. Start from Maxwell's equation for curl H,
$\nabla \times H = J$, and take the divergence of both sides, which must identically be zero

$$\nabla . \nabla \times H = \nabla . J = 0$$

for div curl $F = 0$.

However, we already know this to be wrong, at least for the continuity equation
$\nabla . J = -\partial\rho_V/\partial t \neq 0$ (see section 3.2). It is not acceptable to take $\partial\rho_V/\partial t = 0$
always, although there are situations in which ρ_V and hence $\partial\rho_V/\partial t$ are zero. The
simplest compatible assumption is to write

$$\nabla . \nabla \times H = \nabla . J + \frac{\partial\rho_V}{\partial t} = \nabla . J + \nabla . \frac{\partial D}{\partial t} = 0$$

from which

$$\nabla \times H = J + \frac{\partial D}{\partial t} = J + J_D$$

This is the completed Maxwell third equation.

Now the first term J is the current density and may itself be considered as composed of two terms

$$J = J_f = J(\text{conduction}) + J(\text{convection}) = \sigma E + \rho_V u$$

The second term $\partial D/\partial t$ is the *displacement* current density J_D so that it must have the dimensions A/m^2, as have σE and $\rho_V u$.

Having recognised $\partial D/\partial t$ as being a current density J_D the big achievement of Maxwell was to state that it too, like J_f, could produce a magnetic field.

6.2.1 Displacement current in a flat-plate capacitor

The classic example of displacement J_D and current density I_D is the capacitor. In an a.c. circuit there is conduction current through all components except the capacitor where one assumes the 'charging' current to be $I_e = dQ/dt$. So that for a sine-wave applied voltage v of $V_m \sin(\omega t)$, the capacitor current $I_c = C(dv/dt) = V_m \omega C \cos(\omega t)$, and this is also the 'conduction' current elsewhere in the circuit.

Assuming the capacitor to be a pair of parallel plates of area A and separation d, the capacitance is $A\epsilon_0/d$ farads and the charging (conduction) current I_c becomes $(\omega A \epsilon_0 V_m/d) \cos(\omega t)$. Does this current flow across the dielectric of the capacitor? Certainly it does not flow as conduction current, for there are no free charges (electrons) to carry it. The dielectric may in fact be free space, vacuous, containing nothing (aether?), in which case the displacement current cannot be carried by bound electrons either. Follow Maxwell and accept that $\partial D/\partial t$ may be related to a magnetic field in an equivalent manner to the relationship between a magnetic field and a current density J_D.

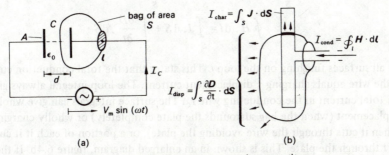

Figure 6.4 *Displacement current in a capacitor.*

A parallel-plate capacitor supplied with a voltage $V_m \sin(\omega t)$ is shown in figure 6.4a. The plate area is A and the separation of the plates is d. Imagine one plate of the capacitor C to be surrounded by an 'open (plastic) bag' surface S. According to

Maxwell there is a 'displacement' current I_D completing the circuit.

$$I_D = \int_S J_D \cdot dS = \int_S \frac{\partial D}{\partial t} \cdot dS$$

We already know from circuit theory that this should be $(\omega A \epsilon_0 V_m / d) \cos(\omega t)$. Let us perform the 'displacement' current calculation for the capacitor using the above equations.

The electric flux density

$$D = \frac{Q}{A} \text{ (ignoring 'fringing')} = \frac{CV_m \sin(\omega t)}{A}$$

$$= \frac{A \epsilon_0}{d} \frac{V_m \sin(\omega t)}{A} = \frac{\epsilon_0}{d} V_m \sin(\omega t)$$

$$\frac{\partial D}{\partial t} = \frac{\omega \epsilon_0 V_m}{d} \cos(\omega t) a_n \quad \text{(normal to plate)}$$

$$I_D = \int_S \frac{\partial D}{\partial t} \cdot dS = \int_S \frac{\omega \epsilon_0 V_m}{d} \cos(\omega t) a_n \cdot dS$$

To carry out this surface integral, now imagine the bag to be of rubber so that it collapses on to the *one* plate. S becomes the area A, only one side carries D, and $I_D = \omega \epsilon_0 V_m \cos(\omega t)(A/d) = I_c$ as before.

However, if this appears unconvincing, we can perform a surface integral over the actual, unusual, bag surface S by invoking Stokes' theorem.

Starting from Maxwell's modified curl H equations

$$\nabla \times H = J + J_D$$

$$\int_S \nabla \times H \cdot dS = \int_S J \, dS + \int_S J_D \cdot dS$$

$$\oint_l H \cdot dl = \int_S J \cdot dS + \int_S \frac{\partial D}{\partial t} \cdot dS$$

for all surfaces finishing on the loop l. This states that the total conduction current in the wire equals charging + displacement current. The loop integral always gives the total current in the conducting wire I_c. The surface integrals can give wholly displacement (when the bag surrounds the plate completely) or wholly charging (when it cuts through the wire avoiding the plate), or a portion of each if it cuts part through the plate. This is shown in an enlarged diagram, figure 6.4b. If the whole plate *is* in the bag as shown in figure 6.4

$$I_D = \int_S \frac{\partial}{\partial t} D \cdot dS = I_c = \frac{\omega \epsilon_0 V_m A}{d} \cos(\omega t)$$

Thus when we take $\oint E \cdot dl$ for the e.m.f. in a circuit containing a plate condenser we can assume that the displacement current path between the plates is in the $E \cdot dl$ loop and in fact it completes the loop.

6.2.2 Displacement current in a concentric capacitor

The concentric capacitor or long unloaded concentric cable brings out the proportioning of current components rather better than the flat-plate capacitor. A similar reasoning is used to that in the previous section and the cylindrical capacitor is shown in figure 6.5. Imagine the bag to cover any length of the inner cable from the far end which is open. The integral over the surface of the bag gives the displacement current which is always equal to the input conduction current given by the loop integral at its mouth.

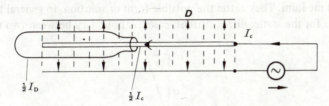

Figure 6.5 *Displacement current in a concentric capacitor.*

At the input end, all the current I_c is carried in as conduction current, and through the bag as displacement current I_D $(= I_c)$; halfway down the line and surface integrals give $\frac{1}{2} I_c$ and $\frac{1}{2} I_D$.

The interesting observation about the concentric cylinder is that the conductivity current measured in the centre core falls progressively from I_c at the input end to zero at the far end. The circular magnetic field strength H produced depends on the core conductivity current and it also falls progressively from $(I_c/2\pi\rho)a_\phi$ at the input end to zero at the far end.

6.3 Maxwell's equations, final forms

We are now in a position to restate them, taking into account the time-varying modifications.

Differential form (fields) Integral form (boundaries)

$$\text{div } D = \nabla \cdot D = \rho_V \quad (1) \qquad \oint_S D \cdot dS = \int_{\text{vol}} \rho_V \, d(\text{vol})$$

$$\text{curl } E = \nabla \times E = -\frac{\partial B}{\partial t} \quad (2) \qquad \oint_l E \cdot dl = -\frac{\partial}{\partial t} \int_S B \cdot dS$$

$$\text{curl } H = \nabla \times H = J + \frac{\partial D}{\partial t} \quad (3) \qquad \oint_l H. \, dl = I + \int_S \frac{\partial D}{\partial t} . \, dS$$

$$\text{div } B = \nabla . B = 0 \qquad\qquad (4) \qquad \oint_S B. \, dS = 0$$

Auxiliary relations

$$D = \epsilon E = \epsilon_0 E + P \qquad J = \sigma E + \rho_V u$$

$$B = \mu H = \mu_0 H + M \qquad F = \rho_V (E + u \times B)$$

6.4 Boundary conditions

These dominate all but one field problem which is wave propagation in an infinite (unbound) medium. They settle the possible form of solution. In general these are the same as for the static, direct current, cases. A boundary between two media is

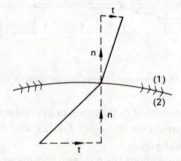

Figure 6.6 *Defining boundary relations.*

shown in figure 6.6 for a general vector F, and this defines the boundary conditions.

General

$$E_{t_1} = E_{t_2}$$

$$H_{t_1} - H_{t_2} = K$$

$$D_{n_1} - D_{n_2} = \rho_S$$

$$B_{n_1} = B_{n_2}$$

Perfect conductor (2) $\sigma_2 = \infty$ Perfect insulator (2) $\sigma_2 = 0$

$E_{t_1} = 0$	$E_2 = 0$	$E_{t_1} = E_{t_2}$
$H_{t_1} = K$	$H_2 = 0$	$H_{t_1} = H_{t_2}$
$D_{n_1} = \rho_S$	$J_2 = 0$	$D_{n_1} - D_{n_2} = \rho_S$
$B_{n_1} = 0$	$K = K$	$B_{n_1} = B_{n_2}$

There are a few noteworthy points of interest.

(i) ρ_S is possible for dielectrics, conductors and insulators.

(ii) For the perfect conductor σ is infinite; the current can be carried on the surface only as K, not as J.

(iii) For the perfect conductor E cannot exist (*vide* Ohm's law) and, since $\nabla \times E = -\partial B/\partial t$, nor can B and H exist. This is not the case for steady currents and static charges.

(iv) The dielectric and magnetic boundary relations involving relative ϵ_r and μ_r values for E_n, D_t and H_n, B_t are usually applied after first finding the other components from the above relations.

It begins to be obvious that the application of boundary state conditions is extremely important.

6.5 Field functions

Sometimes these are derived experimentally. Their range of validity may be checked by reference to Maxwell's equations.

As an example, suppose that a cylindrically symmetrical electromagnet, as shown in figure 6.7, was found to have an air-gap flux density given approximately by $B = [a(4 - \rho^2) \sin(\omega t)]a_z$ tesla; $\omega = 10^8 \times 3$ s^{-1}. Maxwell's equations can be employed as follows.

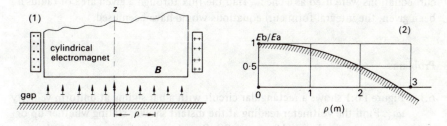

Figure 6.7   *Cylindrically symmetrical electromagnet.*

(i) *E via curl H.* $\nabla \times H = \partial D/\partial t = \epsilon_0(\partial E/\partial t)$; $J = 0$ *in vacuo*. Since $H = B/\mu_0$ we know that only $H_z (\rho,t)$ exists, so that the appropriate term in curl H is

$$-\frac{\partial H_z}{\partial \rho}a_\phi = \left[-\frac{a}{\mu_0}(-2\rho)\sin(\omega t)\right]a_\phi = \epsilon_0\frac{\partial E}{\partial t}$$

Hence

$$E = \left[-\frac{a2\rho}{\mu_0\epsilon_0\omega}\cos(\omega t)\right]a_\phi = E_a \text{ (say)}$$

(ii) *E via curl E.* $\nabla \times E = -\partial B/\partial t = -a\omega(4 - \rho^2) \cos(\omega t)a_z$. In curling E we have to produce an a_z component only and since E will be also cylindrically symmetrical the only appropriate term is $(1/\rho)\{\partial(\rho E_\phi)/\partial\rho\} a_z = \{-a\omega(4 - \rho^2) \times \cos(\omega t)\}a_z$.

Hence

$$\frac{\partial(\rho E_\phi)}{\partial\rho} = -a\omega(4\rho - \rho^3) \cos(\omega t)$$

and finally

$$E = \left\{-a\omega\rho \, \frac{2 - \rho^2}{4} \, \cos(\omega t)\right\} a_\phi = E_b \text{ (say)}$$

To compare the two results, which ought to be identical if the field function is a true Maxwellian field, take the ratio

$$\frac{E_b}{E_a} = \omega^2\mu_0\epsilon_0\left(1 - \frac{\rho^2}{8}\right) = 1 - \frac{\rho^2}{8}$$

The constant $\omega^2\mu_0\epsilon_0$ is found to be unity for the given ω which is fortunate. Computing the ratio E_b/E_a for different values of ρ gives

ρ	0	0·5	1	1·5	2	3	metres
$\|E_b/E_a\|$	1	0·969	0·875	0·719	0·500	−0·0125	

which is shown in (2). Clearly the experimentally derived expression for B has gross errors above about $\rho = 0·5$ m.

In the above, because B was stated in field or point form, the two point-form curl equations were used as a check. Had the flux through a given area of radius ρ been given, the integral-form curl equations would have been used.

Problems

6.1   Figure P6.1 shows a rectangular circuit with a bar sliding at uniform velocity ua_x. Find the voltmeter reading at the instant shown, stating whether up or down scale, for $L = 2$, $M = 1$, $u = 20$, $B_0 = 1$ for the following extensive magnetic flux densities, all in the z direction: (a) $B_0 a_z$, (b) $B_0 \exp(-at)a_z$, (c) $B_0 \exp(\beta x)a_z$, and (d) $B_0 \cos(\omega t - \beta x)a_z$ where $a = 15$, $\beta = 0·1$, $\omega = 20\pi$.

((a) − 20;   (b) +2·23;   (c) −24·4;   (d) −32·1)

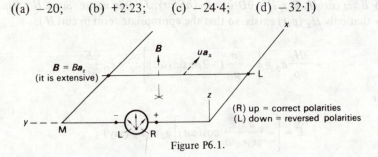

Figure P6.1.

6.2 A magnetic field of flux density B in air is given by $B_0 \exp(bt)a_z$. Assuming a circular path $\rho = a$, calculate the e.m.f. round the path and hence the electric field strength E and electric flux density D. From E find the magnetic field strength H using the curl H equation, and then find B again. Comment.

$(-\frac{1}{2}bB_0a \exp(bt)$, B is not a true Maxwell field)

6.3 Compute the crest value of the displacement current densities for the following: (i) a field in a h.v. cable of $\epsilon_r = 4$ and electric field strength $E = 400 \sin(100\pi t)$ kV/m, (ii) a laser beam of 10^{15} Hz and E of 30 kV/m and (iii) a radio wave of 1 MHz and 1 μV/m.

((i) 4·44 mA/m², (ii) 1·67 x 10⁹ A/m²; (iii) 55·6 pA/m²)

6.4 The magnetic field strength in the air gap of a circularly symmetrical magnet is given by

$$H = H_0 \left\{ 1 - \left(\frac{\rho}{a}\right)^2 \right\} \cos(\omega t)a_z$$

Find the electric field strength E in the air gap via curl E. Test the validity of the expression for H.

$\left(\frac{1}{4}\omega\mu_0 H_0\rho \left\{ 2 - \left(\frac{\rho}{a}\right)^2 \right\} \sin(\omega t)a_\phi; \text{ not a true Maxwellian field} \right)$

6.5 (a) For a coaxial cable radii a, b, length L and permittivity ϵ, evaluate the displacement current I_D across an intermediate surface ($a < \rho < b$) for a sinusoidal applied voltage $v_m \sin(\omega t)$. Show that I_D is independent of ρ and equal to the charging current.
(b) A 1500 m long cable has $a = 1$, $b = 2$ cm, $\epsilon_r = 3·6$. A voltage $10^5(2)^{1/2} \sin(100\pi t)$ is applied at one end. Compute the magnetic field strength H midway in the insulation, at the input end and halfway along the cable.

((b) $204a_\phi$ A/m, $102a_\phi$ A/m)

6.6 A rectangular loop of wire of width w and length h is placed with its plane in the plane of a parallel-wire transmission line with h parallel to the length of the wires. The transmission line wires are a distance a apart and the distance between the length of the loop nearest the line to the nearest wire of the line is b. The current in the line changes at a rate of dI/dt. Calculate the induced e.m.f. in the loop by considering (a) the magnetic flux linkage through the loop and (b) the rate of change of magnetic vector potential.

$\left(\frac{\mu_0 h}{2\pi} \frac{dI}{dt} \ln \left\{ \frac{b(a + b + \omega)}{(a + b)(b + \omega)} \right\} \right)$

6.7   A time-varying magnetic flux density is given by

$$B = B_0 \cos(\omega t)a_z \qquad \text{for } \rho < a$$
$$B = 0 \qquad \text{for } \rho > a$$

Find the induced electric field in both regions.

$$\left(\frac{B_0\rho\omega}{2} \sin(\omega t)a_\phi, \qquad \frac{B_0 a^2 \omega}{2r} \sin(\omega t)a_\phi \right)$$

6.8   A circular conducting plate of radius a, thickness t and conductivity σ is
       rotated with a constant angular velocity ω radians per second at right
       angles to a uniform magnetic field of flux density B. Calculate (a) the
       potential between the edge of the plate and the centre, and (b) the average
       power loss. The conducting plate is now placed at right angles to and stationary
       with respect to an alternating magnetic field of flux density $B_0 \sin(\omega t)$.
       Calculate (c) the total current flowing in the plate, and (d) the average
       power loss.

       ((a) $\omega B_0 a^2/2$;   (b) 0;   (c) $\sigma \omega t B_0 a^2 \cos(\omega t)/4$;
       (d) $\sigma \omega^2 t \pi B_0^2 a^4/16$)

6.9   A charge q is moving along the x axis with velocity v. Consider a circle of
       circumference a in the yz plane whose centre is on the x axis a distance x_1
       from the charge. Calculate the magnetic field strength at a point on the
       circumference by (a) considering the moving charge to be a current element
       and using the Biot–Savart law, and (b) calculating the displacement current
       flowing through the circle and then applying Ampère's law.

6.10   An electron of mass m_0 and charge $-q_e$ starts from rest at $t = 0$ in a region
       in which uniform mutually perpendicular magnetic and electric fields of flux
       density B_0 and electric field strength E_0 exist. Show that the resulting
       equations of motion are

$$-\frac{mE_0}{q_e B_0}\left[\frac{q_e B_0 t}{m} - \sin\left(\frac{-q_e B_0 t}{m} \right) \right] \text{along } E$$

$$-\frac{mE_0}{q_e B_0^2}\left[1 - \cos\left(\frac{-q_e B_0 t}{m} \right) \right] \text{perpendicular to } E \text{ and } B$$

       Show that the electric field seen by an observer in motion with the electron
       has similarly oriented components.

       ($E_0 \sin(-q_e B_0 t/m)$ along E, and $E_0 \cos(-q_e B_0 t/m)$ perpendicular to both
       E and B)

7

Electromagnetic Waves

We are now in a position to examine the applications of Maxwell's equations to free space. In this there are no free charges, no free currents and no magnetic or dielectric material. Most important, however, is that in the first instance there are no boundary conditions to fulfil. Obviously here the conduction current is equal to zero and the solution is the easiest possible with the point form of the equations in use. Maxwell's equations for curl E and curl H indicate a coupling between time-varying magnetic and electric fields. This in turn suggests that whenever a change in electric or magnetic field occurs then a transfer process, consisting of energy flow from the electric to magnetic to electric to magnetic and so on, is initiated. Since the electric and magnetic effects are not confined to the same location, then the transfer of energy in the form of an electromagnetic wave is a logical outcome.

There are two important physical phenomena to note before any mathematical analysis. These are

(i) The progressions indicated by Maxwell's equations, namely

$$\text{curl } E \rightarrow -\mu_0 \frac{\partial}{\partial t} H$$

$$\text{curl } H \rightarrow +\epsilon_0 \frac{\partial}{\partial t} E$$

must take time to propagate.

(ii) With no free charges present div $D = \epsilon_0$ div $E = 0$ and so the electric field flow lines no longer terminate on charges. This must not be interpreted to mean that E does not exist. The time rate of change of E constitutes a displacement current density J_D which is possible in free space.

7.1   Wave motion

Here we need the basic equations for a simple one-dimensional travelling wave.

In figure 7.1, the wave profile moves with a constant velocity u in the z direction and, at time $t = 0$, the wave profile is given by $x = f(z') = f(z)$. At time $t = t$, the wave profile is $x = f(z') = f(z - ut)$. Hence $f(z - ut)$ represents a wave of profile $x = f(z)$ which is *travelling forwards* in the z direction at velocity u. This includes also $f(ut - z)$ which is also forward travelling.

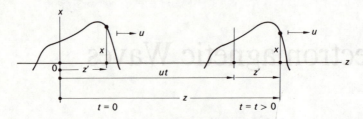

Figure 7.1   *Simple travelling wave.*

By inference $f(z + ut)$, or $f(ut + z)$, is a wave of profile $f(z)$ *travelling backwards*, in the $-z$ direction, at a velocity u. The functions of most practical interest are the sine and cosine functions.

7.2   Free-space wave equations

Maxwell's equations in point form can be used since there are to be no boundary restrictions. These will be focused on the electric field strength E and magnetic field strength H.

$$\text{curl } H = \nabla \times H = \frac{\partial}{\partial t} D = \epsilon_0 \frac{\partial}{\partial t} E$$

$$\text{curl } E = \nabla \times E = -\frac{\partial}{\partial t} B = -\mu_0 \frac{\partial}{\partial t} H$$

$$\text{div } B = \mu_0 \text{ div } H = \mu_0 \nabla . H = 0$$

$$\text{div } D = \epsilon_0 \text{ div } E = \epsilon_0 \nabla . E = 0$$

Now consider the two curl-equations differentiating the first equation with respect to time and curling the second, which is of course with respect to distance.

$$\left.\begin{array}{l} \dfrac{\partial}{\partial t}(\nabla \times H) = \dfrac{\partial}{\partial t}\left(\epsilon_0 \dfrac{\partial}{\partial t} E\right) \\[2ex] \nabla \times \dfrac{\partial}{\partial t} H = \epsilon_0 \dfrac{\partial^2}{\partial t^2} E \end{array}\right\} \text{ from the first}$$

$$\nabla \times (\nabla \times E) = -\mu_0 \nabla \times \frac{\partial}{\partial t} H \left.\begin{array}{c} \\ \\ \\ \\ \end{array}\right\} \text{ from the second}$$

$$\nabla (\nabla . E) - \nabla^2 E = -\mu_0 \nabla \times \frac{\partial}{\partial t} H$$

Because $\nabla . E = 0$ in free space, we can combine the pair to obtain

$$\nabla^2 E = \mu_0 \epsilon_0 \frac{\partial^2}{\partial t^2} E$$

this is the *wave equation* for E, the corresponding one for H is

$$\nabla^2 H = \mu_0 \epsilon_0 \frac{\partial^2}{\partial t^2} H$$

The $\nabla^2 H$ equation has been obtained conversely by curling the first of the equations and time differentiating the second.

These are well-known wave equations and they have their counterparts in many other physical media. As already discussed we expect solutions of the form $f_1(z - ut)$ and $f_2(z + ut)$. It will be easier to show this for particular simple cases, depending on the space components of E, H. We shall often refer to an E wave or an H wave and we hope that readers will accept that this means an electromagnetic wave whose electric or magnetic field variation is given by the particular expression.

7.2.1 *Possible form of solution of the wave equation*

Let us take the very simplest wave to consider. Suppose that we choose z as the direction of propagation and limit E to one component $E = E_x a_x$. E_x is dependent solely on z.

The E-wave equation becomes

$$\nabla^2 E = \nabla^2 E_x a_x + \nabla^2 E_y a_y + \nabla^2 E_z a_z = \frac{\partial^2}{\partial z^2} E_x a_x$$

From this, we need to solve

$$\nabla^2 E = \frac{\partial^2}{\partial z^2} E_x = \mu_0 \epsilon_0 \frac{\partial^2}{\partial t^2} E_x$$

The unit vector a_x has been omitted and a scalar wave equation results.

Let us now test to see if $f_1(z - ut)$ is a solution.

$$\frac{\partial}{\partial z} f_1 = f'(z - ut); \qquad \frac{\partial^2}{\partial z^2} f_1 = f''(z - ut)$$

$$\frac{\partial}{\partial t} f_1 = -uf'(z - ut); \qquad \frac{\partial^2}{\partial t^2} f_1 = +u^2 f''(z - ut)$$

In these the prime $'$ indicates a partial derivative with respect to z. From these equations

$$\frac{\partial^2}{\partial z^2} f_1 = \frac{1}{u^2} \frac{\partial^2}{\partial t^2} f_1$$

and clearly $f_1 = z - ut$ is a solution provided only the *forward* velocity $u = u_0 = 1/(\mu_0\epsilon_0)^{1/2} \approx 3 \times 10^8$ m/s. This is the speed of electromagnetic (including light) waves in free space. Here the symbol u_0 is used to indicate free space, *vide* ϵ_0, μ_0. Some writers use c for the free-space velocity of electromagnetic radiation.

An exactly similar result is obtained for $f_2(z + ut)$ which is a wave travelling *backwards*; it would be at the same speed. Finally the complete solution becomes

$$E = E_x a_x = [f_1(z - ut) + f_2(z + ut)]a_x$$

7.2.2 *Sinusoidal waveforms*

This is the most important type of variation since in practice sinusoidal variations of source and field are often encountered. Also any function with an arbitrary time variation may be expressed as an infinite sum of discrete or continuous frequency spectrum sinusoids, corresponding to periodic or non-periodic variations of the original function. The space distribution of a monochromatic wave (imagined to be a snapshot of the wave at some instant in time) is a sinusoid. If the electric and magnetic field strength vectors E and H are limited to such shapes then it is possible to use the complex algebra of phasors.

Supposing $E_s = |E|[\cos(\omega t) + j\sin(\omega t)] = |E|\exp(j\omega t)$ we need only use the time function $\exp(j\omega t)$ on all vectors throughout and then select the cosine or sine components by taking the real or imaginary parts of the resulting complex expression. Additionally E could represent the maximum value E_m or the root mean square value $E_m/(2)^{1/2}$; accordingly E_s will be either the maximum or r.m.s. value. The s-subscript notation here refers to 'sine wave' or more correctly the '$s = j\omega$' transform of the sine wave. E_s thus denotes a physical quantity which varies sinusoidally in time and is also a vector in space.

The phasor representation of a sinusoidally time-varying quantity is often written as $\bar{F}(x, y, z)$ so that the expression for say a sinusoidally time-varying magnetic field might be written as $H = \text{Re}[\bar{H}(x, y, z)\exp(j\omega t)]$. The phasor representation need not be confined to a vector quantity and a charge distribution ρ which varied in this way would be represented by $\rho = \text{Re}[\rho(x, y, z)\exp(j\omega t)]$. *The phasor representations are complex quantities which are functions of space only.*

Maxwell's equations in phasor form may be written in free space as follows

$$\nabla \times \bar{E} = -j(\omega)\mu_0\bar{H}$$

$$\nabla \times \bar{H} = j(\omega)\epsilon_0\bar{E}$$

$$\nabla \cdot \bar{E} = 0$$

$$\nabla \cdot \bar{H} = 0$$

and the phasor form of the wave equation is

$$\nabla^2 \bar{E} = -\omega^2 \mu_0 \epsilon_0 \bar{E}$$

It is possible to use the phasor representation and to treat the equations as purely spatial variations or to use the s-subscript notation and to retain the time dependence in implicit form. Either course is perfectly acceptable and it is usually a matter of personal preference which method is used in practice. In this text we shall use both approaches.

Now reverting to the single-component wave, $E_s = E_{x_s} a_x$ is a component varying sinusoidally in time only and is an x vector only in space. As before we consider propagation in the z direction.

$$\nabla^2 E_s = \frac{\partial^2}{\partial z^2} E_{x_s} a_x = (-\omega^2 \mu_0 \epsilon_0 E_{x_s}) a_x$$

Now take the two possible solutions, f_1 and f_2, writing

$$E_x = E_{x_1} \exp[j(\omega t - \beta z)] + E_{x_2} \exp[j(\omega t + \beta z)]$$

$$\frac{\partial^2}{\partial z^2} E_x = (-j\beta)^2 E_{x_1} \exp[j(\omega t - \beta z)] + (j\beta)^2 E_{x_2} \exp[j(\omega t + \beta z)]$$

$$\frac{\partial^2}{\partial z^2} E_{x_s} = -\beta^2 E_{x_s} = -\omega^2 \mu_0 \epsilon_0 E_{x_s}$$

For this solution to hold, $\beta^2 = \omega^2 \mu_0 \epsilon_0$ from which

(i) *phase constant* $\beta = \omega/u_0 = \dfrac{2\pi f}{f \lambda_0} = \dfrac{2\pi}{\lambda_0}$ rad/m

(ii) *wavelength* $\lambda_0 = \dfrac{2\pi}{\beta} = \dfrac{u_0}{f}$ m

Here f is the frequency in hertz or cycles per second. The symbol λ has also been used for magnetic flux linkages; from the differences in context, confusion is not likely. The field just described might well be represented by the following expressio for the electric field strength in the x direction E_x

$$E_x = E_{x_1} \cos\{\omega[t - z(\mu_0 \epsilon_0)^{1/2}]\}$$

taking the real part of the complex formulation.

7.2.3 Uniform plane wave

Having found the simple one-component E wave, we ought now to find the corresponding H wave.

The expression for the electric field may be used as follows

$$E_s = E_{x_s} a_x$$

$$\text{curl } E_s = \nabla \times E_s = \frac{\partial}{\partial z} E_{x_s} a_y$$

$$= -\mu_0 \frac{\partial}{\partial t} H_s = -\mu_0 j \omega H_s$$

whence

$$H_s = -\frac{1}{j\omega\mu_0} \frac{\partial}{\partial z} E_{x_s} a_y = H_{y_s} a_y$$

Considering only the forward travelling wave for which the E solution is $E_x = E_{x_1} \exp[j(\omega t - \beta z)]$, the possible H solution is

$$H = -\frac{1}{j\omega\mu_0}(-j\beta)E_{x_1}\exp[j(\omega t - \beta z)]a_y$$

$$= H_y a_y = H_{y_1}\exp[j(\omega t - \beta z)]a_y$$

From this note that E_{x_1}/H_{y_1} has the dimension of impedance (volts per metre/ amperes per metre)

$$z_0 = E_{x_1}/H_{y_1} = \omega\mu_0/\beta = (\mu_0/\epsilon_0)^{1/2} = 120\pi = 377 \ \Omega$$

This is the *intrinsic impedance of free space*. Again the subscript 0 is used, as in ϵ_0, μ_0, u_0, λ_0, therefore z_0. Do not confuse this with the dimension z along the z axis. In some texts the symbol η is used for intrinsic impedance.

We see now that the necessary relation between the electric and magnetic field strengths E and H has been established. That is for single components

$$E_s = E_{x_s}a_x = z_0 H_{y_s}a_x$$

$$H_s = H_{y_s}a_y = \frac{1}{z_0}E_{x_s}a_y$$

Choosing the sine-wave component for simplicity we could rewrite these *in extenso* as

$$E = E_{x_1}\sin(\omega t - \beta z)a_x$$

$$H = H_{y_1}\sin(\omega t - \beta z)a_y$$

Note that E has values only in the x direction, and H has values only in the y direction. *They are in time phase with each other* and we can represent them pictorially as shown in figure 7.2. This simple wave is as follows

(i) *Plane*—at any point on a plane z = constant the values of E are the same; so too are all the values of H.

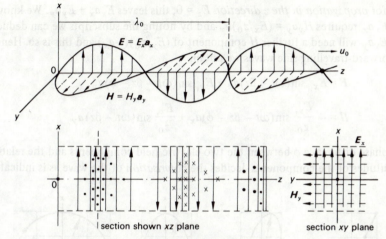

Figure 7.2 *Uniform plane wave.*

(ii) *Transverse*—because everywhere the electric and magnetic field strengths E and H are transverse or orthogonal to the direction of propagation.

(iii) *Polarised* — because E has values only in one direction (x) and H has values only in one direction (y here). The wave is said to be polarised in the x direction meaning that is the *direction of the only* electric field strength it possesses.

The uniform plane wave is a TEM wave, that is transverse electromagnetic, but cannot exist physically because it stretches to infinity and would represent therefore an infinite energy. If, however, a wave is considered some distance from its source then often the approximation to a uniform plane wave is applicable. Not all waves will be TEM waves. The *microwaves* propagated in *waveguides* will be shown to be either TE (transverse electric) or TM (transverse magnetic) waves.

7.2.4 Plane wave with two components

We can now ask what would be the result of starting with an E wave of more than one component. Say for instance

$$E = E_x a_x + E_y a_y + E_z a_z$$

and again restrict E_x, E_y, E_z to be functions of z, t only.

Hence

$$\frac{\partial E}{\partial x} = \frac{\partial E}{\partial y} = 0$$

and

$$\nabla \cdot E = \frac{\partial E_z}{\partial z} = 0$$

thus for *propagation in the z direction* $E_z = 0$; this leaves $E_x a_x + E_y a_y$. We know that $E_x a_x$ requires $H_y a_y = (E_x/z_0)a_y$ and by noting the subscripts we can deduce that $E_y a_y$ will need a further H component of $(E_y/z_0)(-a_x)$ and this is so. Hence for forward-travelling sine waves we obtain

$$E = E_{x_1} \sin(\omega t - \beta z)a_x + E_{y_1} \sin(\omega t - \beta z - \phi)a_y$$

$$H = -\frac{E_{y_1}}{z_0} \sin(\omega t - \beta z - \phi)a_x + \frac{E_{x_1}}{z_0} \sin(\omega t - \beta z)a_y$$

The phase difference ϕ between the two electric field components and the relative magnitudes of the components decides the *polarisation* of the wave as is indicated

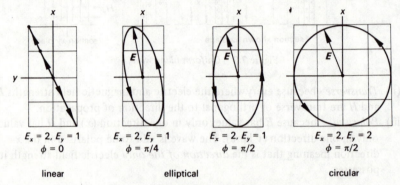

$E_x = 2, E_y = 1$ $E_x = 2, E_y = 1$ $E_x = 2, E_y = 1$ $E_x = 2, E_y = 2$
$\phi = 0$ $\phi = \pi/4$ $\phi = \pi/2$ $\phi = \pi/2$

linear   elliptical   circular

Figure 7.3   *Polarisation of plane wave.*

in figure 7.3. Four loci are illustrated for the electric field at a fixed value of z. The path traced by the tip of the resultant field vector at the fixed value of z is a straight line for $\phi = 0$, an ellipse for $E_x \neq E_y, \phi \neq 0$ and a circle for $E_x = E_y$, $\phi = \pi/2$.

7.3 Plane waves in homogeneous loss-free media

Suppose we now are concerned with waves in homogeneous, isotropic, loss-free media of dielectric constant ϵ_r and relative permeability μ_r, both of which are constant. Exactly the same analysis will apply as before with the following alterations

$$u = \frac{1}{(\epsilon\mu)^{1/2}} = \frac{u_0}{(\epsilon_r \mu_r)^{1/2}} = \frac{f\lambda_0}{(\epsilon_r \mu_r)^{1/2}}$$

$$\lambda = \frac{\lambda_0}{(\epsilon_r \mu_r)^{1/2}}; \quad \beta = \frac{2\pi}{\lambda} = \frac{2\pi}{\lambda_0}(\epsilon_r \mu_r)^{1/2}$$

$$Z_0 = \left(\frac{\mu}{\epsilon}\right)^{1/2} = z_0\left(\frac{\mu_r}{\epsilon_r}\right)^{1/2}$$

This is called variously the intrinsic, wave or surge impedance. As an example consider pure water, with $\epsilon_r = 81$ and $\mu_r = 1$, sustaining electromagnetic waves at $f = 3 \times 10^8$ Hz.

Wavelength $\lambda_0 = 1$ m for free space

$$\text{velocity } u_0 = 3 \times 10^8 \text{ m/s}$$

Here

$$\lambda = \frac{\lambda_0}{(81 \times 1)^{1/2}} = \frac{\lambda_0}{9} = \frac{1}{9}\text{m} = 0{\cdot}111 \text{ m}$$

$$\text{Velocity } u = \frac{u_0}{9} = 0{\cdot}333 \times 10^8 \text{ m/s}$$

$$\text{Phase constant } \beta = \frac{2\pi}{\lambda} = 18\pi = 56{\cdot}5 \text{ rad/m}$$

$$\text{Intrinsic impedance } Z_0 = \frac{z_0}{9} = \frac{120\pi}{9} = 41{\cdot}9 \ \Omega$$

7.4 Poynting's vector $\mathscr{P}$

This is due to J. H. Poynting (Birmingham, 1884). It is concerned with the magnitude and direction of the density of power flow in an electromagnetic field. We are heading towards $E \times H$; V/m $\times$ A/m = W/m^2 which is the correct dimension for power density.

Proceed from the curl H equation

$$\nabla \times H = J + \frac{\partial}{\partial t} D$$

$$E . \nabla \times H = E . J + \epsilon E . \frac{\partial}{\partial t} E$$

Using the identity

$$\nabla . (E \times H) = H . \nabla \times E - E . \nabla \times H$$

$$- \nabla . (E \times H) + H . \nabla \times E = J . E + \epsilon E . \frac{\partial}{\partial t} E$$

$$- \nabla . (E \times H) = J . E + \epsilon E . \frac{\partial}{\partial t} E - H . \nabla \times E$$

use the curl E equation

$$- \nabla . (E \times H) = J . E + \tfrac{1}{2}\epsilon \frac{\partial}{\partial t} E^2 - H\left(-\frac{\partial}{\partial t} B\right)$$

$$= J . E + \tfrac{1}{2}\epsilon \frac{\partial}{\partial t} E^2 + \tfrac{1}{2}\mu \frac{\partial}{\partial t} H^2$$

Now integrate all the terms through the volume considered

$$\int_{\text{vol}} - \nabla . (E \times H) \, d(\text{vol}) = \int_{\text{vol}} J . E \, d(\text{vol}) + \int_{\text{vol}} \frac{1}{2} \epsilon \frac{\partial}{\partial t} E^2 \, d(\text{vol})$$

$$+ \int_{\text{vol}} \frac{1}{2} \mu \frac{\partial}{\partial t} H^2 \, d(\text{vol})$$

Use the divergence theorem on the left to change the volume integral to a surface integral

$$\oint_S \{-(E \times H) . dS\} = \int_{\text{vol}} J . E \, d(\text{vol}) + \frac{\partial}{\partial t} \int_{\text{vol}} (\frac{1}{2} \epsilon E^2 + \frac{1}{2} \mu H^2) \, d(\text{vol})$$

total instantaneous _ instantaneous ohmic + rate of increase of
power *into* vol ⎯ loss in vol   (electric + magnetic
                                             stored energy in vol)

Hence

$$\text{total instantaneous power out of volume} = \oint_S (E \times H) . dS$$

and

$$\text{instantaneous power density } \mathscr{P} = E \times H \text{ (W/m}^2 \text{ of surface)}$$

This is Poynting's theorem, the vector $\mathscr{P}$ being the Poynting ('pointing') vector; the direction of the vector is defined in figure 7.4a. This agrees with the uniform plane wave considered, for E_x and H_y give z propagation.

In applying this theorem we must ensure that E and H are causally related, that is both are simultaneously from the same source. In the arrangement in figure 7.4b the electric field is from a charged capacitor and the magnetic field from a permanent magnet; $\mathscr{P} = 0$.

The Poynting vector is along the direction of propagation of an electromagnetic wave and can help to resolve between left- and right-handed polarisation. If the rotation of the electric field strength vector E is considered, as in figure 7.4, then if

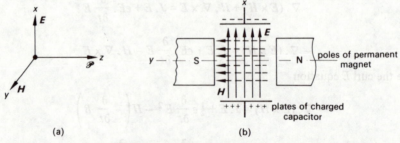

(a)   (b)

Figure 7.4   *Illustrating Poynting's theorem.* (a) *Direction of $E \times H$.* (b) *Situation in which $E \times H = 0$.*

E is viewed along the direction of propagation (the direction of the Poynting vector) a clockwise rotation corresponds to right-handed polarisation while for left-handed polarisation the rotation is in the opposite sense. Here the wave is moving away from the observer and for a wave moving towards the observer the rotation of the electric field strength vector is viewed in the opposite direction to the direction of the Poynting vector. In this case a clockwise rotation represents left-handed polarisation and an anti-clockwise rotation represents right-handed polar-isation. Unless the direction of the Poynting vector is given ambiguities may arise in the sense of the polarisation.

The definitions of the sense of polarisation just considered are in accordance with the usual practice in radio engineering and constitute the definition of right- and left-handed polarisation adopted by the Institute of Electrical and Electronic Engineers. However, it is still possible for ambiguities to arise since the definition of right- and left-handed polarisation is accepted to be in the opposite sense in the field of classical optics. It is most important to pay particular attention to the meaning of right- and left-handed polarisation (or clockwise and anti-clockwise) in magneto-optical materials and laser communications, where according to background the meaning of the sense of polarisation may be different. All ambiguities may be removed if the definition is made with respect to the radiation received by a right- or left-handed helical beam antenna and the radio-engineering definition is used. A right-handed helical beam antenna receives, or radiates, right-handed polarised waves only.

7.4.1 Stored energy-density components

For a wave in a loss-free medium there are, if the electric field strength is E and the magnetic field strength is H

$$\frac{\epsilon E^2}{2} \quad \text{and} \quad \frac{\mu H^2}{2} \text{ J/m}^3$$

in the electric and magnetic fields respectively. The simple plane wave we are considering has

$$E = E_{x_1} \sin(\omega t - \beta z)a_x \qquad H = H_{y_1} \sin(\omega t - \beta z)a_y$$

The time average of the square of a sine wave of maximum value v_m is $\frac{1}{2}v_m^2$ (the r.m.s. value is well known as $v_m/(2)^{1/2}$). Therefore the time averages of the energy densities are

$$W_E = \frac{\epsilon E^2_{x_1}}{4} \qquad W_M = \frac{\mu H^2_{y_1}}{4} \quad \text{J m}^{-3}$$

Consider

$$\frac{W_E}{W_M} = \frac{\epsilon}{\mu}\left(\frac{E_{x_1}}{H_{y_1}}\right)^2 = \frac{1}{Z_0^2}Z_0^2 = 1$$

This means that the electromagnetic wave energy is carried equally in the electric and magnetic fields. There is no interchange of energy between the two fields for they, and the instantaneous energy densities, are co-phased in time.

7.4.2 *The Poynting vector in a power cable*

This is an interesting example and figure 7.5a shows a section of a long, one-core, concentric cable. First suppose the core and sheath to be loss-free, that is $\sigma = \infty$, and a direct current I is being carried at voltage V.

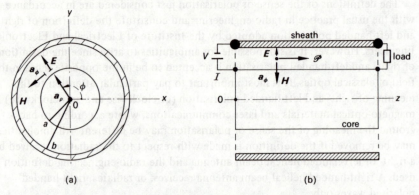

Figure 7.5 *Poynting vector $\mathscr{P}$ in loss-free coaxial (concentric) cable on load.*
(a) *Cable cross-section.* (b) *Cable terminated by load.*

For the concentric cable, charged q C/m, the electric field strength E and magnetic field strength H are given by

$$H = \frac{I}{2\pi\rho}\, a_\phi \qquad E = \frac{q}{2\pi\epsilon\rho}\, a_\rho$$

$$V = \frac{q}{2\pi\epsilon \ln(b/a)} \qquad E = \frac{V}{\rho \ln(b/a)}\, a_\rho$$

$$\mathscr{P} = E \times H = \frac{VI}{2\pi\rho^2 \ln(b/a)}\, a_z \text{ W/m}^2$$

This is the power density flow along the dielectric towards the load, figure 7.5b, so that the total power flow will be found over the circular end with ρ varying between the limits b, a. $dS = 2\pi\rho\, d\rho a_z$

$$P(\text{watts}) = \int_S \mathscr{P}.\,dS = \frac{VI}{2\pi \ln(b/a)} \int_a^b \frac{2\pi\rho\, d\rho}{\rho^2}\, a_z . a_z$$

$$= \frac{VI}{\ln(b/a)} \ln \rho \Big|_a^b = VI \text{ as expected}$$

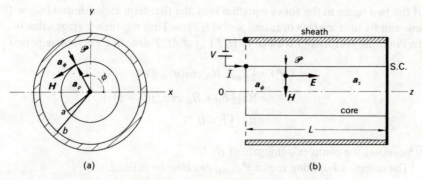

Figure 7.6   *Poynting vector 𝒫 in concentric cable with core loss only.* (a) *Cable cross-section.* (b) *Cable terminated by short circuit.*

The surprising thing is that this appears to flow down the dielectric between core and sheath which serve only to guide 𝒫 but *not* to carry it!

Next suppose the core has resistance, but not so the sheath, and let them be short circuited at the far end. At the surface of the core, see figure 7.6a, b

$$H = \frac{I}{2\pi a} a_\phi \text{ A/m}; \qquad E = \frac{V}{L} a_z \text{ V/m}$$

The latter relation comes from observing that the whole voltage V is lost in moving down the length L, presumably uniformly. Hence $\mathscr{P} = E \times H = (V/L)a_z \times (I/2\pi a)a_\phi = -(VI/2\pi aL)a_\rho$ which means that the power flow is *into* the core where it is being dissipated as $I^2 R$ losses. The total power flow *into* the core is

$$P = - \oint_S \mathscr{P}. \, dS = \frac{VI}{2\pi aL} \, a_\rho . (2\pi aL)a_\rho = VI$$

This is as expected and equal to $I^2 R$ in the core. It could otherwise be computed by $\int_{\text{vol}} J. E \, d(\text{vol})$ in the core.

7.4.3   Complex Poynting vector and the phasor notation

The previous results on the Poynting vector must be modified when we consider sinusoidally time-varying fields. Suppose such a field is characterised by complex field vectors

$$\bar{E} = E_0 \exp(j\theta) \qquad \text{and} \qquad \bar{H} = H_0 \exp(j\phi)$$

then the instantaneous value of the Poynting vector is

$$\mathscr{P}_{\text{inst}} = E \times H = \text{Re}[\bar{E} \exp(j\omega t)] \times \text{Re}[\bar{H} \exp(j\omega t)]$$

$$= \frac{E_0 \times H_0}{2} [\cos(2\omega t + \theta + \phi) + \cos(\theta - \phi)]$$

Of the two terms in the above equation only the first term varies sinusoidally with time and its time average over one period is zero. Thus the time-average value of the Poynting vector $\langle \mathscr{P} \rangle$ is equal to $(I/T) \int_0^T \mathscr{P}\,\mathrm{d}t$, T denoting a complete period.

$$\langle \mathscr{P} \rangle = \tfrac{1}{2} E_0 \times H_0 \cos(\theta - \phi)$$
$$= \mathrm{Re}[\tfrac{1}{2} E_0 \times H_0 \exp\{\mathrm{j}(\theta - \phi)\}]$$
$$= \mathrm{Re}(\tfrac{1}{2}\bar{E} \times \bar{H}^*)$$

$\bar{H}^*$ denotes the complex conjugate of $\bar{H}$.

The complex Poynting vector $\mathscr{P}_{\mathrm{comp}}$ can now be defined.

$$\mathscr{P}_{\mathrm{comp}} = \tfrac{1}{2}\bar{E} \times \bar{H}^*$$

so that we have

$$\langle \mathscr{P} \rangle = \mathrm{Re}(\bar{\mathscr{P}}_{\mathrm{comp}}) = \tfrac{1}{2}\,\mathrm{Re}(\bar{E} \times \bar{H}^*)$$

We now proceed to calculate the complex power flow and we consider a volume V bound by a closed surface S.

$$\oint_S \mathscr{P}_{\mathrm{comp}} \cdot \mathrm{d}S = \oint (\bar{E} \times \bar{H}^*) \cdot \mathrm{d}S = \tfrac{1}{2} \int_{\mathrm{vol}} \nabla \cdot (\bar{E} \times \bar{H}^*)\,\mathrm{d}(\mathrm{vol})$$

Now using Maxwell's equations in the complex form

$$\nabla \times \bar{E} = -\mathrm{j}\omega\mu_0\bar{H}$$
$$\nabla \times \bar{H} = \bar{J} + \mathrm{j}\omega\epsilon_0\bar{E}$$

one can write

$$\nabla \cdot (\bar{E} \times \bar{H}^*) = \bar{H}^* \cdot (-\mathrm{j}\omega\mu_0\bar{H}) - \bar{E} \cdot (\bar{J} \times \mathrm{j}\omega\epsilon_0\bar{E})^*$$
$$= -\mathrm{j}\omega\mu_0\bar{H}^* \cdot \bar{H} - \bar{E} \cdot \bar{J}^* + \mathrm{j}\omega\epsilon_0\bar{E} \cdot \bar{E}^*$$

The time-average stored energy density in the electric field is given by

$$\langle W_{\mathrm{E}} \rangle = \langle \tfrac{1}{2}\epsilon_0 E^2 \rangle$$
$$= \langle \tfrac{1}{2}\epsilon_0 |E_0|^2 \cos^2(\omega t + \theta) \rangle$$
$$= \langle \tfrac{1}{4}\epsilon_0 |E_0|^2 + \tfrac{1}{4}\epsilon_0 |E_0|^2 \cos^2(\omega t + \theta) \rangle$$
$$= \tfrac{1}{4}\epsilon_0 |E_0|^2 = \tfrac{1}{4}\epsilon_0 E_0 \exp(\mathrm{j}\theta) \cdot E_0 \exp(-\mathrm{j}\theta)$$
$$= \tfrac{1}{4}\epsilon_0 \bar{E} \cdot \bar{E}^*$$

A similar calculation for the time-average stored energy density in the magnetic field $\langle W_{\mathrm{M}} \rangle$ yields

$$\langle W_{\mathrm{M}} \rangle = \tfrac{1}{4}\mu_0 \bar{H} \cdot \bar{H}^*$$

The time average of the power density is

$$\langle E . J \rangle = \text{Re}(\tfrac{1}{2}\bar{E} \times \bar{J}*)$$

and the complex power density for the current flow is $\tfrac{1}{2}\bar{E} . \bar{J}*$ and is written $\bar{p}_d$.
The expression for $\nabla . (\bar{E} \times \bar{H}*)$ now becomes

$$\nabla . (\bar{E} \times \bar{H}*) = -2\bar{p}_d - 4j\omega(\langle W_H \rangle - \langle W_E \rangle)$$

and

$$\oint \mathscr{P}_{\text{comp}} . dS = - \int_{\text{vol}} \bar{p}_d \ d(\text{vol}) - j2\omega \int_{\text{vol}} (\langle W_H \rangle - \langle W_E \rangle) \ d(\text{vol})$$

This is the *complex Poynting theorem* and is used for complex fields. Separating
the real and imaginary parts

$$\int_{\text{vol}} \langle p_d \rangle \ d(\text{vol}) = - \oint_S \langle \mathscr{P} \rangle . dS \qquad\qquad \text{real}$$

$$\int_{\text{vol}} (\langle W_H \rangle - \langle W_E \rangle) \ d(\text{vol}) = -\frac{1}{2\omega} \left[\text{Im}\left(\oint_S \mathscr{P}. dS \right) + \text{Im}\left(\int_{\text{vol}} p_d \ d(\text{vol}) \right) \right]$$

$$\text{imaginary}$$

Consider a wave propagating in the z direction such that the electric field is

$$E = E_0 \exp(j\omega t)a_x$$

and the field in the x direction is

$$E_x = E_0 \exp\left[j\omega \left(\frac{t-z}{u} \right) \right]$$

Thus for the y direction

$$-j\omega\mu_0 H_y = \frac{\partial E_x}{\partial z}$$

$$H_y = \frac{\epsilon}{\mu} E_x$$

The power carried by a plane wave may be calculated using the complex Poynting
theorem.

$$\langle \mathscr{P} \rangle = \mathscr{P}_z a_z = \tfrac{1}{2} \text{Re}(\bar{E} \times \bar{H}*)$$

$$= \tfrac{1}{2}\left(\frac{\epsilon}{\mu} \right)^{1/2} E_x{}^2 a_z$$

This result indicates that the time-average power flow is the same at all points and
is in the direction of propagation of the waves—a result which should be obvious
since there can be no power loss in free space.

7.5 Plane waves in a medium having loss

The loss could be from poor conduction in the medium. Presumably it might also arise from dielectric loss or even from eddy current and hysteresis in the medium if it is magnetic. Here we consider only the conductivity of the σ medium.

The curl equations will now become, for the magnetic field strength H and electric field strength E

$$\nabla \times H = J + \frac{\partial}{\partial t} D = \sigma E + \epsilon \frac{\partial}{\partial t} E$$

$$\nabla \times E = -\frac{\partial}{\partial t} B = -\mu \frac{\partial}{\partial t} H$$

Next suppose we are dealing only with sine waves; these are

$$\nabla \times H = \sigma E + j\omega\epsilon E = (\sigma + j\omega\epsilon) E$$

$$\nabla \times E = -\mu j\omega H$$

Consider curl curl E expanded to (grad div-lap)E

$$\nabla \times \nabla \times E = \nabla(\nabla . E) - \nabla^2 E = -\mu j\omega \; \nabla \times H = -\mu j\omega(\alpha + j\omega\epsilon)E$$

Hence assuming there are no free charges in the medium, so that div $D = 0$ and $\nabla . E = 0$, we have

$$\nabla^2 E = +j\omega\mu(\sigma + j\omega\epsilon) E = \gamma^2 E \qquad \text{(say)}$$

A similar process obtains for H so that the wave equations are

$$\nabla^2 E = \gamma^2 E \qquad \text{and} \qquad \nabla^2 H = \gamma^2 H$$

Now restricting attention to the simple x-polarised plane wave we have

$$E = E_x a_x$$

where E_x is $f(z, t)$, so $\nabla^2 E$ becomes $(\partial^2/\partial z^2) E = \gamma^2 E$. A solution to this is $E = E_{x_1} \exp(-\gamma z) a_x$. Note here that we have taken only the $-\gamma$ possibility which corresponds to the forward travelling wave; $+\gamma$ for the backward wave has been ignored.

Bringing back the time element we write

$$E = E_{x_1} \exp(j\omega t) \exp(-\gamma z) a_x$$

The form this takes will depend on the propagation constant γ. Suppose we write $\gamma = \alpha + j\beta$ then

$$E = E_{x_1} \exp(-\alpha z) \exp\{j(\omega t - \beta z)\}] a_x = E_x$$

from which we see that a new factor has been brought in: α, the attenuation constant, whose units are nepers per metre. The phase constant β (radians per metre) we have had before. Thus, as the E wave progresses in the z direction, it not only

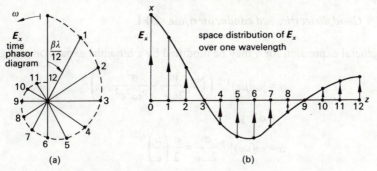

Figure 7.7 *Space distribution of E for a wave in a lossy medium.* (a) *Phase variation.* (b) *Attenuation over one wavelength.*

changes its phase according to $\exp(-j\beta z)$, but it also attenuates according to $\exp(-\alpha z)$. The variations of the electric field are shown in figure 7.7. The two diagrams show how E_x changes in phase and attenuation over one wavelength. In the diagrams for one-twelfth of a wavelength, $\beta\lambda/12 \equiv 30°$ and $\exp(-\alpha\lambda/12)$ is about 0·9. Carefully note that E_x is always in the $0x$ direction. There will be analogous diagrams for H_y which now is not in time phase with E_x.

For propagation with loss, the constant γ is

$$\gamma = (j\omega\mu(\sigma + j\omega\epsilon))^{1/2} = j\omega(\mu\epsilon)^{1/2}\left(1 - j\frac{\sigma}{\omega\epsilon}\right)^{1/2} = \alpha + j\beta$$

How this is to be interpreted depends greatly on the size of the factor $\sigma/\omega\epsilon$ which we already know to be the ratio of (conduction current/displacement current) in the lossy dielectric medium.

As regards wave impedance Z_0 this can be obtained from the two solutions to the wave equations

$$E = E_x \exp(j\omega t) \exp(-\gamma z)a_x$$

$$H = \frac{\gamma}{j\omega\mu} E_x \exp(j\omega t)\exp(-\gamma z)a_y$$

From these equations, and ignoring unit vectors a_x, a_y

$$Z_0 = \frac{E}{H} = \frac{j\omega\mu}{\gamma} = \frac{(j\omega\mu)^{1/2}}{(\sigma + j\omega\epsilon)^{1/2}} = \frac{(\mu\epsilon)^{1/2}}{(1 - j(\sigma/\omega\epsilon))^{1/2}}$$

This is now clearly complex and its phase angle will influence H which can no longer be in time phase with E but will lag behind.

Now interpret these results for two main classes of media. The first is a relatively good dielectric and bad conductor; conversely the second is a good conductor and bad dielectric.

APPLIED ELECTROMAGNETICS

7.5.1 *Good dielectric, bad conductor* $\sigma/\omega\epsilon < 0\cdot 1$

The general expression for γ may be modified by a binomial expansion giving

$$\gamma = j\omega(\mu\epsilon)^{1/2}\left(1 - \frac{j\sigma}{2\omega\epsilon} + \frac{\sigma^2}{8\omega^2\epsilon^2}\cdots\right)$$

$$= \alpha + j\beta$$

$$\alpha = \omega(\mu\epsilon)^{1/2}\frac{\sigma}{2\omega\epsilon} = \frac{\sigma}{2}\left(\frac{\mu}{\epsilon}\right)^{1/2}$$

$$\beta = \omega(\mu\epsilon)^{1/2}\left(1 + \frac{\sigma^2}{8\omega^2\epsilon^2}\right)$$

$$u = \frac{\omega}{\beta} = \left[(\mu\epsilon)^{1/2}\left(1 + \frac{\sigma^2}{8\omega^2\epsilon^2}\right)\right]^{1/2}$$

$$= \frac{1}{(\mu\epsilon)^{1/2}} = \frac{u_0}{(\epsilon_r\mu_r)^{1/2}}$$

The wave velocity u is reduced below u_0. The wave impedance Z_0 is given by

$$Z_0 = \left(\frac{\mu}{\epsilon}\right)^{1/2}\left(1 + \frac{j\sigma}{2\omega\epsilon}\right)$$

and will have a small positive phase angle so that H lags slightly behind E.

7.5.2 *Good conductor, poor dielectric* $\sigma/\omega\epsilon > 10$

The approximation is made by taking

$$\left(1 - \frac{j\sigma}{\omega\epsilon}\right)^{1/2} = \left(\frac{-j\sigma}{\omega\epsilon}\right)^{1/2}$$

$$= \left(\frac{\sigma}{\omega\epsilon}\right)^{1/2}\exp\left(\frac{-j\pi}{4}\right)$$

$$\gamma = j\omega(\mu\epsilon)^{1/2}\left(\frac{\sigma}{\omega\epsilon}\right)^{1/2}\exp\left(\frac{-j\pi}{4}\right)$$

$$= (\omega\mu\sigma)^{1/2}\exp\left(\frac{+j\pi}{4}\right)$$

$$= \left(\frac{\omega\mu\sigma}{2}\right)^{1/2}(1 + j1)$$

$$\alpha = \beta = \left(\frac{\omega\mu\sigma}{2}\right)^{1/2}$$

$$u = \frac{\omega}{\beta} = \left(\frac{2\omega}{\mu\sigma}\right)^{1/2}$$

$$Z_0 = \left(\frac{j\omega\mu}{\sigma}\right)^{1/2} = \left(\frac{\omega\mu}{\sigma}\right)^{1/2} \exp\left(\frac{j\pi}{4}\right)$$

$$= \left(\frac{\mu}{\epsilon}\right)^{1/2} \left(\frac{\omega\epsilon}{\sigma}\right)^{1/2} \exp\left(\frac{j\pi}{4}\right)$$

$$= \left(\frac{\mu}{\epsilon}\right)^{1/2} \left(\frac{\omega\epsilon}{2\sigma}\right)^{1/2} (1 + j1)$$

$$= \left(\frac{\omega\mu}{2\sigma}\right)^{1/2} (1 + j1)$$

The wave impedance has a $45°$ phase angle; by the same angle H lags behind E in time phase. This is a surprising result. The impedance Z_0 has a magnitude less than z_0 (377 ohms).

As the conductivity σ increases, the attenuation α and the wave velocity u both decrease.

7.5.3 Dissipation factor D

The dissipation factor D is $\sigma/\omega\epsilon$ and also the ratio of conduction current density to displacement current density, J_C/J_D. It is dominant in determining the wave-transmission characteristics of the medium. It is also related to the loss angle δ in a capacitor having the same medium as dielectric.

The diagram, figure 7.8, shows how α and β depend on dissipation factor D. It is interesting to remark that the propagation characteristics of the ground, though dependent upon ω, fall somewhere between the two extremes of bad and good conductors.

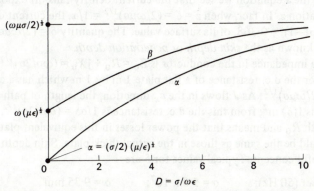

Figure 7.8   *Effect of dissipation factor D on propagation components.*

7.5.4 Skin effect

It is well known that in a good conductor high-frequency currents are mostly confined to the surface layers of the conductor. To show this consider propagation in a medium such as copper. The displacement current everywhere will be negligible. The conductor is shown in figure 7.9 which indicates how the surface field penetrates the conductor.

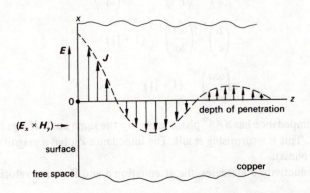

Figure 7.9 *Skin effect; depth of wave penetration in conductor.*

Suppose a surface electric field of strength E exists

$$E = E_{x_1} \cos(\omega t) a_x$$

In the conductor it will be $E = E_{x_1} \exp(-\alpha z) \cos(\omega t - \beta z) a_x$ where $\alpha = \beta = (\omega \mu \sigma / 2)^{1/2}$. In the conductor everywhere $J = \sigma E$ so that

$$J = E_{x_1} \exp(-\alpha z) \cos(\omega t - \beta z) a_x$$

$$H_y = E_{x_1} \left(\frac{\sigma}{\omega \mu} \right)^{1/2} \exp(-\alpha z) \cos\left(\omega t - \beta z - \frac{\pi}{4} \right) a_y$$

Considering the J equation we see that the current density falls off exponentially with penetration z. In fact, when $z = \delta = (2/\omega \mu \sigma)^{1/2} = 1/\alpha$, the current density has fallen to $\exp(-1)$, $= 0.368$, of its surface value. The quantity $\delta = (2/\omega \mu \sigma)^{1/2} = 1/\alpha = 1/\beta$ is known as the *skin depth* or *penetration depth*.

The wave impedance in the conductor is $Z_0 = R_0 + jX_0 = (\omega \mu / 2\sigma)^{1/2}(1 + j1)$. Now consider the d.c. resistance of a flat plate 1 m by 1 m which has a conductivity depth $\delta = (2/\omega \mu \sigma)^{1/2}$. As J flows in the a_x direction, the length of path is 1 m and the section is $1(\delta)$ m^2; from this the d.c. resistance is $1/\sigma \delta = (\omega \mu / 2\sigma)^{1/2}$. This is identical with R_0 and means that the power losses in this equivalent plate of depth δ would be the same as those in the actual thick plate. Skin depth δ is clearly a useful concept. Typical values for δ are

(i) copper (50 Hz); $\sigma = 5.8 \times 10^7$; $\delta = 9.35$ mm
(ii) silver (10^{10} Hz); $\sigma = 6.6 \times 10^7$; $\delta = 6.41 \times 10^{-4}$ mm

These figures explain why solid conductors for power frequency are rarely more than 2 cm diameter; and why an evaporated film of silver (10^{-4} inch) on glass is an excellent conductor at microwave frequencies.

The total current flowing in 1 m width of very thick plate can be obtained as follows

$$I = \int_0^\infty J \, dz = \int_0^\infty \sigma E_{x_1} \exp(j\omega t) \exp(-\gamma z) \, dz = \frac{\sigma}{\gamma} E_{x_1} \exp(j\omega t)$$

$$= \frac{\sigma\delta}{(2)^{1/2}} E_{x_1} \cos\left(\omega t - \frac{\pi}{4}\right) \text{amperes}$$

By carrying out the same integral between limits of z from 0 to δ, it can be shown that 85·9 per cent of the total current just derived flows in a skin depth δ of the plate.

The skin depth δ found for a flat plate can be used for approximate results for cylindrical conductors in which radius $a \gg \delta$; that means, at 50 Hz, for wires which have diameters exceeding about 2 cm. For high frequencies when $a \gg \delta$, $R_{hf} = 1/2\pi a\sigma\delta$ ohms per metre length; this is the resistance of a hollow tube of radius a and thin walls of depth δ.

7.5.5 Flat-plate transmission line

Flow down a concentric cable has already been considered. It is instructive to consider the flat-plate transmission line in the light of skin depth.

For transmission into a thick copper plate, the wave impedance Z_0 is $(\omega\mu/2\sigma)^{1/2}(1 + j1)$; at 50 Hz this is about $1\cdot845(1 + j1)\,\mu\Omega$. The attenuation and phase shift constants α, β are very high, each being about 107. The wave velocity u is very low, 2·94 m/s; the wavelength is small, 5·86 cm.

A pair of thick flat plates acting as a transmission line is shown in figure 7.10, each plate being long in the y direction.

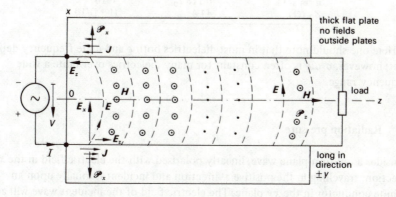

Figure 7.10 *Thick flat-plate transmission line with loss.*

The electric field strength E at the plate is bent as shown; at the plate it has two components

$$E = E_t + E_n = E_z + E_x$$

The magnetic field strength H due to the current in the plate is all in the y direction. The Poynting vector has two components

$$\mathscr{P}_z = E_x \times H_y \qquad \text{(towards the load)}$$

$$\mathscr{P}_x = E_z \times H_y \qquad \text{(to the plates)}$$

The first component (the major one) carries the load power; the wave velocity for this is as usual. The second component carries the $I^2 R$ power for the plate loss; the wave velocity inside the plate is low. Of course if σ were infinite $\mathscr{P}_x$ would be zero. The plates appear to serve as 'guides' for the electromagnetic wave carrying the main power to the load.

7.5.6 Propagation in sea water

Sea water, like the ground, behaves either like a conductor or like an insulator depending on the frequency of the electromagnetic waves. The properties of sea water are $\sigma = 4$ S/m, $\epsilon_r = 72$, $\mu_r = 1$, thus the frequency for which $\omega \epsilon = \sigma$ is 10^9 Hz. We shall compare the properties of 10^{10} Hz and 10^4 Hz, at which frequencies sea water is a good dielectric and a good conductor respectively. See table 7.1.

Table 7.1

	10^{10} Hz	10^4 Hz
α (neper m^{-1})	89·0	0·5
β (rad m^{-1})	1777	0·5
δ (m)	$1\cdot12 \times 10^{-2}$	2·0
λ (m)	$3\cdot55 \times 10^{-3}$	12·6
λ_0 (m)	3×10^{-2}	3×10^4
u (m s^{-1})	$0\cdot118 u_0$	$4\pi 10^4$
Z_0 (Ω)	44·4	$(1 + j)/10$

Here one should note that in most dielectrics both ϵ and σ are frequency-dependent; however, $\sigma/\omega\epsilon$ is often constant for many dielectrics over quite a wide frequency range.

7.6 Radiation pressure

Consider a uniform plane wave, linearly polarised with the electric field in the x direction, travelling in the positive z direction and incident normally upon an infinite conductor in the xy plane. The electric field of the incident wave will act upon the electrons in the conductor and cause a current which in turn interacts with the magnetic field of the wave to produce a force. By the Lorentz equation, section 5.1, if ρ_v is the volume charge density due to free electrons within the conductor, the force acting on the conductor per unit volume is

$$F = \rho_v(E + u \times B)$$

Now

$$\text{div } D = \rho_v, \qquad \text{curl } H = J + \dot{D} \qquad \text{and} \qquad J = \rho_v u$$

so

$$F = E \text{ div } D + (\text{curl } H) \times B - \dot{D} \times B$$

and

$$\frac{d}{dt}(D \times B) = \dot{D} \times B + D \times \dot{B} = \dot{D} \times B - D \times \text{curl } E$$

$$F = \epsilon_0 E \text{ div } E + \mu_0(\text{curl } H) \times H - \mu_0\epsilon_0 \frac{\partial}{\partial t}(E \times H) - \epsilon_0 E \times \text{curl } E$$

$$= -\frac{\mu_0}{2} \frac{\partial}{\partial z}(H_y^2) a_z - \frac{\epsilon_0}{2} \frac{\partial}{\partial z}(E_x^2) a_z - \mu_0\epsilon_0 \frac{\partial}{\partial t}(E \times H)$$

Now the time-average power flow is constant so $(\partial/\partial t)(E \times H) = 0$ and the force in the z direction is

$$F_z = -\frac{\partial}{\partial z}\left(\frac{\epsilon_0}{2} E_x^2 + \frac{\mu_0}{2} H_y^2\right)$$

The term inside the bracket will be recognised as the stored energy, see section 7.4, so the force equals the rate of change with distance of the stored energy. The force acts in the positive z direction and in the conductor energy decreases with distance owing to the absorption of energy.

In a uniform plane wave $E_x/H_y = (\mu_0/\epsilon_0)^{1/2}$, the energy density is $\epsilon_0 E_x^2$, $\mathscr{P} = E_x H_y a_z = u_0\epsilon_0 E_x^2 a_z$, and

$$F_z = -\frac{1}{u_0} \frac{\partial \mathscr{P}}{\partial z}$$

and when a wave is incident upon a conductor it is absorbed within a small distance δz of the surface. The Poynting vector rapidly decays from $\mathscr{P} = E_x H_y a_z$ at the surface to zero at a distance δz inside the conductor. The force per unit area in the z direction is $\mathscr{P}/u_0$ and this is *the radiation pressure*. In a uniform plane wave

$$\mathscr{P}_{av} = u_0\epsilon_0 E_{rms}^2 a_z = \frac{u_0\epsilon_0}{2} E_0^2 a_z = 2 \cdot 65 \times 10^{-3} E_{rms} a_z \text{ W/m}^2$$

and the average force per unit area $F_{av} = 2\mathscr{P}_{av}/u_0$. The Poynting vector for sunlight is $1 \cdot 4$ kW/m^2 at the surface of the earth; thus the radiation pressure on a

conductor here would be 10^{-5} N/m^2. The absorption of the waves is equivalent to a
change in momentum and the rate of change of momentum produces a force. The
flux of momentum is equal to $\mathscr{P}_{av}/u_0$ which equals the product of momentum
density and wave velocity. This equivalence between radiation and particle type
behaviour may be expressed by means of Einstein's relationship between energy $\mathscr{E}$
and angular frequency ω, $\mathscr{E} = h\omega/2\pi$ where h is Planck's constant.

Problems

7.1   (a) In free space, a plane wave has $E = 0\cdot2 \sin(\omega t - 0\cdot5z)a_x$. Find (i) ω,
(ii) H and (iii) z_0.
(b) By measurement in free space, E has only an x component. The distance
between positive maxima is 5 m in the y direction and 3 m in the z direction.
Find (i) a unit vector specifying the direction of propagation and (ii) the
frequency.

((a) (i) $1\cdot5 \times 10^8$,   (ii) $5\cdot3 \times 10^4 \sin(\omega t - \beta z)a_y$,   (iii) 377 Ω;

(b) (i) $(0a_x + 3a_y + 5a_z)/(34)^{1/2}$,   (ii) $1\cdot168 \times 10^8$ Hz)

7.2   The electric field strength E for a plane wave in a dielectric propagating in
the z direction is given by $E = 4 \cos(10^8 t - z)a_x + 2 \sin(10^8 t - z)a_y$.
     Find (i) λ, f, ϵ_r, (ii) H and (iii) plot E for $z = 0$ as t changes from 0 to
$T = 1/f$.

((i) 2π, $10^8/2\pi$, $\epsilon_r = 9$;
(ii) $(1/20\pi)[-\sin(\omega t - z)a_x + 2\cos(\omega t - z)a_y])$

7.3   (a) Voltage breakdown in air at standard temperature and pressure is
30 kV/cm. What power density in MW/cm^2 can be safely carried by a uniform plane wave (of sine form) allowing 2:1 breakdown safety factor?
(b) A uniform wave propagating in the z direction has a crest amplitude of
2 V/m at $z = 0$. The medium is a lossy dielectric for which $\alpha = 0\cdot01$ and
$Z_0 = 200(1 + j(\approx0))$. Find the average power density at (i) $z = 0$ and
(ii) $z = 100$.

((a) $0\cdot298$ MW/cm^2;   (b) (i) 10 mW/m^2,   (ii) $1\cdot355$ mTa)

7.4   The electric and magnetic fields produced by a radiating source at the coordinate origin in free space are given in spherical coordinates by the electric and
magnetic field strength vectors

$$E = V \frac{\sin \theta}{r} \cos(\omega t - \beta r)a_\theta \text{ V/m}^{-1}$$

$$H = V \frac{\sin \theta}{z_0 r} \cos(\omega t - \beta r)a_\phi \text{ A/m}^{-1}$$

For a spherical surface, radius R, centred on the origin, find expressions for (a) the time-average power density at any point on its surface, and (b) the total steady power propagated through the surface.

((a) $(V^2 \sin^2 \theta / 2z_0 R^2) a_r$;   (b) $(4\pi/3)(V^2/z_0)$)

7.5 (a) The region $z > 0$ is copper of $\sigma = 5\cdot73 \times 10^7$ S/m. At the surface is a tangential electric field strength E of $2 \times 10^{-3} \cos(10^4 t) a_x$. Find E, J and H on a plane at $z = 3\delta$.
(b) Find $\alpha, \beta, u, \lambda$ and Z_0 for a ferrite carrying electromagnetic waves at 10^{10} Hz, given $\sigma = 10^{-2}$ S/m, $\epsilon_r = 9$, $\mu_r = 4$.

((a) $10^{-4} \cos(10^4 t - 3) a_x$, $5730 \cos(10^4 t - 3) a_x$,
$6\cdot73 \cos(10^4 t - 3\cdot785) a_y$;
(b) $1\cdot26$, 1260, $0\cdot5 \times 10^8$,   5 mm, $251(1 + j0\cdot001)$)

7.6 The intrinsic impedance of free space is 377 Ω and a resistive sheet having this resistance is termed a space cloth. Show that the conductivity of such a cloth is $377d$ S/m where d is the thickness of the sheet.

A uniform plane wave with an electric field strength 10 V/m is incident normally upon an infinite space cloth. Calculate the value of the electric field strength in the transmitted and reflected wave.

What would happen if a perfectly conducting sheet was placed a distance $\lambda/4$ behind the space cloth?

($6\cdot67$ V/m, $-3\cdot3$ V/m)

7.7 The electric field strength E with respect to the origin is given by

$$E = \frac{1}{r} \{E_0 \sin \theta \cos(\omega t - \omega(\mu_0 \epsilon_0)^{1/2} r)\} a_\theta$$

in the spherical polar coordinate system. Find expressions for the magnetic flux density and the power flow over a spherical surface centred on the origin.

$$\left(\frac{E_0(\mu_0 \epsilon_0)^{1/2}}{r} \sin \theta \cos \{\omega t - \omega(\mu_0 \epsilon_0)^{1/2} r\} a_\phi, \right.$$

$$\left. \frac{8\pi(\mu_0 \epsilon_0)^{1/2} E_0{}^2 \cos^2 \{\omega t - \omega(\mu_0 \epsilon_0)^{1/2} r\}}{3\mu_0} \right)$$

7.8 A plasma is a gaseous medium containing positive ions and electrons. Assuming the positive ions are immobile and that collisions within the plasma may be neglected show that for sinusoidally time-varying fields, characterised by a phasor electric field strength $\bar{E}$, the wave equation becomes

$$\nabla^2 \bar{E} = -\omega^2 \mu_0 \epsilon_0 \left(1 - \frac{N q_e^2}{m_0 \omega^2 \epsilon_0}\right) \bar{E}$$

where N is the electron density and m_0 the mass of the electron.
What other assumptions are made?

7.9   An electron beam has a velocity v and a charge density ρ_V. If the electrons have mass m_0 and charge $-q_e$ show that Poynting's theorem becomes

$$\nabla \cdot (E \times H) + E \cdot \dot{D} + H \cdot \dot{B} - \frac{m_0}{q_e} J(v \cdot v) - \frac{\partial}{\partial t}\left(\frac{m_0}{2q_e} \rho_V(v \cdot v)\right)$$

Discuss the physical meaning of the last two terms.

7.10 An oscillating dipole with an electric dipole moment $p = p_0 \exp(j\omega t)$ is situated at the origin.
Derive expressions for the electric and magnetic field strengths at the point r, θ, ϕ and the average value of the Poynting vector.

8

Field Problems—Non-exact Solutions

We turn now to consider the solution of field problems with specified boundary conditions. Only static or steady-state problems will be treated and time-varying fields will not be included.

In general, engineering problems are of three dimensions; but there are very many which are basically two-dimensional, that is those in which the field state does not vary along one of the three space-coordinates. We shall confine our attention to these two-dimensional field problems, remarking that with the advent of the digital computer there are a growing number of techniques for dealing with three-dimensional problems. A brief survey of two-dimensional fields only will be possible.

The two point-form differential equations which will concern us are, in terms of the electric potential and cartesian coordinates

$$\nabla^2 V = \frac{\partial^2 V}{\partial x^2} + \frac{\partial^2 V}{\partial y^2} + \frac{\partial^2 V}{\partial z^2} = 0 \qquad \text{Laplace equation}$$

$$\nabla^2 V = \frac{\partial^2 V}{\partial x^2} + \frac{\partial^2 V}{\partial y^2} + \frac{\partial^2 V}{\partial z^2} = -\frac{\rho_V}{\epsilon} \qquad \text{Poisson equation}$$

Equations similar to these arise in many other physical systems, some of which are listed in table 8.1 in terms of the corresponding potential and flux functions.

Table 8.1

	Potential function	Flux function
Electrostatics	electric potential	electric flux line
Conduction	voltage	current
Magnetics	magnetic potential	magnetic flux line
Fluids	velocity potential	stream line
Heat	isothermals	heat flow line
Gravity	gravitational potential	force line
Elasticity	strain energy	stress line
Elastic membranes	deflection	gradient

151

Any field problem concerning particular boundaries has an exact analogue complete with boundaries in any other of these fields. It is therefore not surprising that experimental *analogues* are employed to find solutions to problems in fields where experimentation is difficult.

The four main methods of solving a particular field problem with known boundary conditions are as follows.

(i) Free-hand sketching, with an accuracy of 10 per cent possible.

(ii) Numerical approximation employing iteration, relaxation and, most extensively nowadays, the digital computer. Any desirable degree of accuracy can be obtained, depending upon the skill, patience and determination of the programmer!

(iii) Mathematical analysis employing images, conjugate functions, conformal transformation, power series, cylindrical and spherical harmonics. These are restricted to boundaries definable mathematically.

(iv) Experimental analogues using the electrolytic tank, rubber membranes, soap film, Hele–Shaw fluid-flow equipment, photo-elasticity and more recently conducting paper (Teledeltos), with an accuracy of 2 to 5 per cent.

We shall deal here only with the first three methods. The use of Teledeltos paper will have been undertaken in laboratory work and excellent field-plotting equipment is commercially available. The other analogues tend to be of less importance nowadays though they can often be illuminating.

To start with, the use of conformal transformations and conjugate functions will be explained since these bring out very readily many of the field properties needed in the other methods, particularly in free-hand sketching.

8.1   Two-dimensional fields: basic principles

Here we shall give attention to two-dimensional fields which satisfy Laplace's equation $\nabla^2 V = 0$.

Let us now consider the complex variable $z = x + jy$. Let

$$w = F(z) = F(x + jy)$$

Now $F(z)$ can be expanded into its real and imaginary parts and we write

$$w = u + jv = F(z) = F(x + jy) = f(x, y) + jg(x, y)$$

If $F(z)$ is analytic (that is $\partial F/\partial x$, $\partial F/\partial y$ both exist and are continuous) then it can be shown that the Cauchy–Riemann conditions must hold. They are

$$\frac{\partial u}{\partial x} = \frac{\partial v}{\partial y}$$

Cauchy–Riemann conditions

$$\frac{\partial u}{\partial y} = -\frac{\partial v}{\partial x}$$

Now take the second-order partial derivatives

$$\frac{\partial^2 u}{\partial x^2} = \frac{\partial^2 v}{\partial x\,\partial y} \qquad \frac{\partial^2 u}{\partial x\,\partial y} = \frac{\partial^2 v}{\partial y^2}$$

$$\frac{\partial^2 u}{\partial y^2} = -\frac{\partial^2 v}{\partial x\,\partial y} \qquad \frac{\partial^2 u}{\partial x\,\partial y} = -\frac{\partial^2 v}{\partial x^2}$$

Hence

$$\frac{\partial^2 u}{\partial x^2} + \frac{\partial^2 u}{\partial y^2} = 0 \qquad \text{and} \qquad \frac{\partial^2 v}{\partial x^2} + \frac{\partial^2 v}{\partial y^2} = 0$$

It follows that both u and v are solutions to the two-dimensional Laplace field equation. They are known as *conjugate functions*.

8.1.1   The condition of orthogonality

As before we let $u = f(x, y)$ and $v = g(x, y)$. Suppose we consider $u = f(x, y) =$ constant A; A is real. This must represent a simple curve in the xy plane and it will have a slope dy/dx. Along the curve $u = A$

$$\Delta u = \frac{\partial u}{\partial x}\Delta x + \frac{\partial u}{\partial y}\Delta y = 0$$

Hence

$$\frac{\Delta y}{\Delta x} \rightarrow \left(\frac{dy}{dx}\right)_A = -\frac{\partial u}{\partial x}\left(\frac{\partial u}{\partial y}\right)^{-1}$$

In the same way if we consider $v = g(xy) =$ constant B, then on such a plane curve

$$\frac{\Delta y}{\Delta x} \rightarrow \left(\frac{dy}{dx}\right)_B = -\frac{\partial v}{\partial u}\left(\frac{\partial v}{\partial y}\right)^{-1}$$

Hence using the Cauchy–Riemann conditions

$$\left(\frac{dy}{dx}\right)_A \left(\frac{dy}{dx}\right)_B = \frac{\partial u}{\partial x}\frac{\partial v}{\partial x}\left(\frac{\partial u}{\partial y}\frac{\partial v}{\partial y}\right)^{-1} = -1$$

This means that the $u = A$, $v = B$ curves must *intersect at right angles*, so they are orthogonal. This is a great aid in free-hand mapping.

Hence the two families of curves

$$u = A_0, A_1, A_2, A_3, \ldots; \qquad v = B_0, B_1, B_2, B_3, \ldots$$

form a mesh or pattern of chequers. If the increments $A_1\text{-}A_0, A_2\text{-}A_1, A_3\text{-}A_2, \ldots$ are all equal, and the increments $B_1\text{-}B_0, B_2\text{-}B_1, B_3\text{-}B_2, \ldots$ are likewise equal and so chosen that the mean dimensions in any one mesh $b = l$, then the map is of *curvilinear squares*. This is illustrated in figure 8.1 which is of a two-dimensional field where we have $v =$ constant as the equipotentials and $u =$ constant as the flux lines. The roles can readily be interchanged.

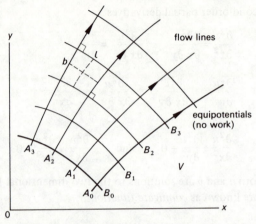

Figure 8.1   *Two-dimensional field of equipotentials and flow lines.*

8.1.2   *Example involving w = exp (x + jy)*

To illustrate the above procedures, take $w = \exp(x + jy)$ and follow in detail

$$w = \exp(x + jy) = \exp x \,(\exp jy) = \exp x \,(\cos y + j \sin y)$$

$$= \exp x \cos y + j \exp x \sin y = u + jv$$

$$\frac{\partial u}{\partial x} = \exp x \cos y; \qquad \frac{\partial v}{\partial y} = \exp x \cos y$$

$$\frac{\partial u}{\partial y} = - \exp x \sin y; \qquad \frac{\partial v}{\partial x} = \exp x \sin y$$

The last two pairs of equations agree with the Cauchy–Riemann conditions. For the second-order partial derivatives

$$\frac{\partial^2 u}{\partial x^2} = \exp x \cos y; \qquad \frac{\partial^2 v}{\partial y^2} = - \exp x \sin y$$

$$\frac{\partial^2 u}{\partial y^2} = - \exp x \cos y; \qquad \frac{\partial^2 v}{\partial x^2} = \exp x \sin y$$

$$\frac{\partial^2 u}{\partial x^2} + \frac{\partial^2 u}{\partial y^2} = 0; \qquad \frac{\partial^2 v}{\partial x^2} + \frac{\partial^2 v}{\partial y^2} = 0 \qquad \text{Laplace equations}$$

From these, $u = \exp x \cos y$ and $v = \exp x \sin y$ are each solutions of Laplace equations.

For equipotentials: $\qquad u = \exp x \cos y = A,\ \left(\dfrac{dy}{dx}\right)_A = \dfrac{\cos y}{\sin y}$

For flow lines: $\qquad v = \exp x \sin y = B,\ \left(\dfrac{dy}{dx}\right)_B = -\dfrac{\sin y}{\cos y}$

Hence $(dy/dx)_A (dy/dx)_B = -1$ and the two families of curves are orthogonal.

8.1.3 Construction of conjugate functions

If we know either u or v, then v or u can be found using the Cauchy-Riemann conditions in the following manner. Suppose we know $u = f(xy)$ and we want to find $v = g(xy)$. We have

$$\frac{\partial u}{\partial x} = \frac{\partial v}{\partial y} \qquad \text{and} \qquad \frac{\partial u}{\partial y} = -\frac{\partial v}{\partial x}$$

From the first

$$v = \frac{\partial f}{\partial x} \, \partial y + C(x) = g(xy)$$

From the second

$$v = \frac{\partial f}{\partial y} \, \partial x + D(y) = g(xy)$$

The last two equations must be identical.

Note that, when carrying out the partial integrals, $C(x)$ may be a constant or a function of x. Similarly $D(y)$ may be either a constant or a function of y. The two solutions must be identically equal, *vide* the uniqueness theorem, and this enables v to be determined completely. The uniqueness theorem may be stated as follows.

If at each and every point of the boundaries of a Laplacian region $\nabla^2 V = 0$, either V or $\partial V/\partial n$ is known, n being the normal, then there is one unique function which satisfies both field and boundary conditions.

This will be assumed to be true and the reader is referred to Hayt (1967, p. 182) for its proof.

By way of illustration, suppose we are given the flux function $u = x^3 - 3xy^2$ and the conjugate potential function v is required. First check that the given u does satisfy Laplace's equation

$$\frac{\partial u}{\partial x} = 3x^2 - 3y^2; \qquad \frac{\partial^2 u}{\partial x^2} = 6x$$

$$\frac{\partial u}{\partial y} = -6xy; \qquad \frac{\partial^2 u}{\partial y^2} = -6x$$

Summing the last two equations to zero, Laplace's equation is satisfied.

$$\frac{\partial v}{\partial y} = 3x^2 - 3y^2; \qquad v = \int (3x^2 - 3y^2) \, \partial y + A(x)$$

$$= 3x^2 y - y^3 + A(x)$$

$$-\frac{\partial v}{\partial x} = -6xy; \qquad v = \int 6xy \, \partial x + B(y)$$

$$= 3x^2 y + B(y)$$

The last two equations must be identically equal and thus, to make these identical, $A(x) = 0$ and $B(y) = -y^3$ and finally $v = 3x^2y - y^3$, giving $w = u + jv = (x^3 - 3xy^2) + j(3x^2y - y^3)$.

The process is not always as simple as in this example.

8.2 Current flow in a thin conducting-sheet

This is important for solution by experimental analogue using, for example, Teledeltos paper.

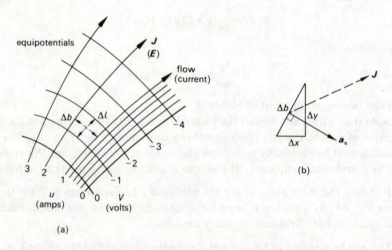

Figure 8.2 *Current flow in a conducting sheet.*

The current flow is shown in figure 8.2a and flux lines and equipotentials are indicated. The flux lines will be orthogonal to the equipotentials. We measure current u in amperes from some convenient zero boundary; likewise potential in volts from a convenient orthogonal boundary. Average voltage gradient is

$$E = -\operatorname{grad} V = -\frac{\partial V}{\partial l} = -\lim\left(\frac{\Delta V}{\Delta l}\right)$$

Let thickness be t and it is often convenient to consider $t = 1$ unit. Also notice here that for convenience the general potential function v has been identified with the electric potential $V(v = V)$.

Let J be the current density in the sheet. Its direction is given by the flow lines and we can write

$$J = \sigma E = -\sigma \operatorname{grad} V = -\sigma \, \nabla V = -\sigma\left(\frac{\partial V}{\partial x}\,a_x + \frac{\partial V}{\partial y}\,a_y\right)$$

Now consider

$$\text{div } \mathbf{J} = -\sigma \nabla . \nabla V = -\sigma \nabla^2 V = -\sigma \left(\frac{\partial^2 V}{\partial x^2} + \frac{\partial^2 V}{\partial y^2} \right)$$

But div $\mathbf{J} = 0$; in the region between the boundaries there are no current sources or sinks so finally

$$\frac{\partial^2 V}{\partial x^2} + \frac{\partial^2 V}{\partial y^2} = 0 \qquad \text{Laplace's equation in two dimensions}$$

The following important points should be noted.

(i) The electric field strength E is the negative potential gradient or $E =$ $-$ grad V thus $V = - \int \mathbf{E} \cdot d\mathbf{l}$. This is taken from some convenient boundary $V = 0$ going *down flow lines*. It is a *scalar* amount of work in volts.

(ii) u is the integral of the current density $+ \int \mathbf{J} \cdot db$ taken from some convenient boundary $u = 0$ going *across the flow lines*; this assumes thickness $t = 1$. It is likewise a *scalar* amount of current, hence it is in amperes.

(iii) As lines of constant u and constant v cut each other orthogonally, u and v must be conjugate functions of the complex variable $z = x + \mathrm{j}y$ as before

$$w = u + \mathrm{j}v = F(z) = F(x + \mathrm{j}y)$$

u is the flow or flux function and $v (= V)$ is the potential function, and as v satisfies Laplace's equation so must u

$$\nabla^2 u = \frac{\partial^2 u}{\partial x^2} + \frac{\partial^2 u}{\partial y^2} = 0 \qquad \text{Laplace's equation for } u$$

(iv) Laplace's equation for u can be independently established as follows. Consider the diagram figure 8.2b.

$$\Delta b = \Delta x a_x + \Delta y a_y; \qquad \mathbf{J} = J_x a_x + J_y a_y$$

$$a_\mathrm{n} = \text{unit vector normal to } \Delta b = \frac{\Delta y a_x - \Delta x a_y}{\Delta b}$$

$$\Delta u = (\mathbf{J} . a_\mathrm{n}) \Delta b = (J_x a_x + J_y a_y) . (\Delta y a_x - \Delta y a_y) = J_x \, \Delta y - J_y \, \Delta x$$

But also $\Delta u = (\partial u / \partial x) \Delta x + (\partial u / \partial y) \Delta y$ hence $J_x = \partial u / \partial y$ and $J_y = - \partial u / \partial x$.

$$\mathbf{J} = \sigma \mathbf{E} = -\sigma \left(\frac{\partial V}{\partial x} a_x + \frac{\partial V}{\partial y} a_y \right) = \partial u / \partial y \, a_x - \partial u / \partial x \, a_y$$

Thus $\partial u/\partial x = \sigma\, \partial V/\partial y$ and $\partial u/\partial y = -\sigma\, \partial V/\partial x$ which are the Cauchy–Riemann conditions with $\sigma = 1$. Proceeding

$$\frac{\partial^2 u}{\partial x^2} = \sigma\,\frac{\partial^2 V}{\partial x\,\partial y}$$

$$\frac{\partial^2 u}{\partial y^2} = -\sigma\,\frac{\partial^2 V}{\partial x\,\partial y}$$

Summing

$$\frac{\partial^2 u}{\partial x^2} + \frac{\partial^2 u}{\partial y^2} = 0$$

(v) The roles of u and v can be interchanged by a simple exchange of boundary conditions. Thus every transformation solves *two* flow problems. For example, see figure 8.3a,b which illustrates how the flow lines and potential functions may be interchanged in a particular problem.

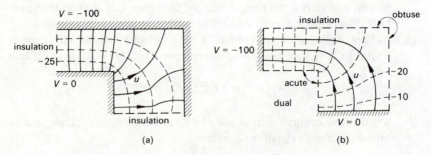

Figure 8.3   *Interchanging roles between flow and potential functions.*

(vi) It is not always necessary to determine both u and v; often if the electric field strength is required it is best to use the potential function V. Certainly this is the easiest to determine experimentally.

8.3   Free-hand mapping

About 10 per cent accuracy can be obtained by sketching free-hand the Laplacian region using curvilinear squares ($l = b$). The following guides may be helpful in making an acceptable field plot, though in fact the construction of a good useful field map is an art in itself.

(i) A conducting boundary is an equipotential line (surface) but an insulating boundary is not.

(ii) Flow lines u meet the equipotential boundaries orthogonally, and similarly cross the equipotentials.

(iii) Equipotentials crowd into boundary corners which have acute angles exterior to the field, and they move away from corners which have obtuse exterior angles.

(iv) Draw the boundary details on graph paper but use transparent paper over it on which to make a preliminary plot.

(v) Use another transparent paper over the first to improve the plot and further ones to make successive improvements. This is better than one paper and an eraser.

(vi) Divide the known potential difference between boundaries into a convenient number of equipotentials, 2, 4, 8, etc.

(vii) Start sketching in the equipotentials v between the boundaries from a region of known (uniform) distribution if there should be one.

(viii) Start the flow lines u similarly from a known (uniform) region, keeping them orthogonal to the v lines and as far as possible with $l = b$. A fraction of a curvisquare may be left finally.

(ix) In the second and subsequent sketches, re-adjust the equipotentials and flow lines to improve the mutual orthogonality and squareness until acceptable.

(x) In regions of low field intensity $(-\nabla v)$ there will be figures often of more than four sides. To check the correctness of the plot, subdivide these regions into curvisquares of $\frac{1}{2}$ (or $\frac{1}{4}$ or $\frac{1}{8}$) size, remembering that as v is subdivided so too must u.

(xi) An improvement in the field pattern may often be suggested by viewing the map upside down or sideways.

(xii) Techniques do exist for dealing with (a) regions of two or more dielectrics and (b) regions having sources in the field. However, these will not be considered here.

8.3.1 Example–cylinder and earthed plate

Suppose we wish to find the conductance (in siemens per metre) between an extensive earthed plate and a long parallel cylinder. This system has already been discussed in section 3.6.3 and here we choose the radius of the cylinder b to be 1 unit and the distance from the plane c to be 3 units.

A half-section of the two-dimensional field pattern is shown in figure 8.4 which is the result of two rough sketches.

There are clearly several cells which are far from good curvisquares and further sketches could profitably be made. From the sketch

$$G = \sigma \, \frac{\text{No. of cells in parallel}}{\text{No. of cells in series}}$$

$$= \sigma \, \frac{(2 \times 7 \cdot 3)}{4} = 3 \cdot 65 \sigma \text{ S/m}$$

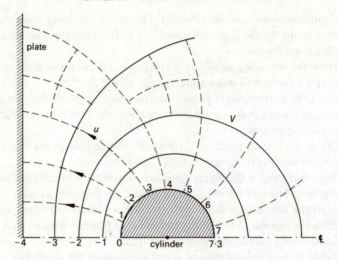

Figure 8.4   *Free-hand mapping; conductance from plate to cylinder.*

The capacitance analogue of this problem has already been solved analytically. We obtained

$$C = \frac{2\pi\epsilon}{\ln\{[c + (c^2 - b^2)^{1/2}]/b\}}$$

Hence

$$G = \frac{2\pi\epsilon}{\ln\{[c + (c^2 - b^2)^{1/2}]/b\}}$$

$$= \frac{2\pi\epsilon}{\ln\{[3 + (9 - 1)^{1/2}]/1\}} = 3\cdot56\sigma$$

This is an excellent check on the result from the free-hand sketch; the accuracy is better than that usually obtained!

8.4   Field correlation with analogues

Suppose that a curvisquare map has been made as shown in the previous example. In going from one potential boundary to the other there will in general be

$$s = \text{cells in series}; \qquad p = \text{cells in parallel}$$

Through each of these cells there will be the same flux (current) and across each cell there will be the same difference of potential (volts). Also $l = b$ so that $l/b = 1$ and $t = 1$.

8.4.1 Conduction field (Teledeltos)

Suppose we take V_1 as the voltage step between any two equipotentials, and I_1 the current flow between any pair of adjacent flow lines. Total potential difference $V = sV_1$; total current $I = pI_1$; σ is the conductivity of material.

$$\text{Conductance per cell} = \frac{I_1}{V_1} = \frac{\sigma \text{ area}}{\text{length}} = \frac{\sigma bt}{l} = \sigma$$

$$\text{Total conductance } G = \frac{pI_1}{sV_1} = \frac{I}{V} = \sigma \frac{p}{s}$$

$$= \sigma \frac{\text{cells in parallel}}{\text{cells in series}}$$

The analogue map found with Teledeltos paper is simply to obtain the ratio p/s. Having found this, the conductance G in the real situation is obtained by using σ of the actual material. Note that only ratio p/s is needed and the cell size does not strictly enter into the estimate. However, a cell size small enough for accuracy is needed.

8.4.2 Thermal field

This is treated in the same way as the conduction field. Let T_1 be the temperature difference between adjacent isothermals; H_1 the heat flow between adjacent flow lines (watts). The total temperature difference $T = sT_1$; the total heat flow $H = pH_1$. If σ_{th} is the thermal conductivity of actual material, then

$$\text{Thermal conductance per cell} = \frac{H_1}{T_1} = \sigma_{th} \frac{bt}{l} = \sigma_{th}$$

$$\text{Total thermal conductance } G_{th} = \frac{H}{T} = \frac{pH_1}{sT_1} = \sigma_{th} \frac{p}{s}$$

This is in thermal siemens per metre thickness.

8.4.3 Electric field

Here we are concerned with capacitance, electric field and electric displacement. Let V_1 be the voltage step between adjacent equipotentials v; $\psi_1 = Q_1$ the electric flux between adjacent flow lines u. The total potential difference $V = sV_1$; total flux = total charge = $p\psi_1 = pQ_1$. If ϵ is the dielectric constant ($\epsilon_r \epsilon_0$) of the actual material

$$\text{Capacitance per cell} = \epsilon \frac{bt}{l} = \epsilon = \frac{Q_1}{V_1}$$

$$\text{Total capacitance } C = \frac{Q}{V} = \frac{p}{s} \frac{Q_1}{V_1} = \epsilon \frac{p}{s} \text{ F/m}$$

The following points should be noted.

(i) If the resistance and thermal resistance, etc., are required, the reciprocals have to be taken, that is

$$R = \frac{1}{G} = \frac{1}{\sigma} \frac{\text{cells in series}}{\text{cells in parallel}}$$

(ii) Regions of high current density and high voltage gradient occur where the cells are smallest. An estimate can be made of the voltage gradient and the flux density by measuring the dimension $l = b$.

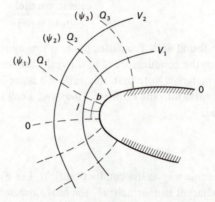

Figure 8.5   *Regions of high potential gradient and flux density.*

Thus for the electric field shown in figure 8.5

$$E = -\text{grad } V \approx -\frac{V_1}{l}$$

$$D = \frac{\text{flux}}{\text{area}} = \frac{\psi_1}{bt} = \frac{Q_1}{b} = \epsilon E$$

This also gives us the electric charge density ρ_S on the surface of the conductor, for $\rho_S = D_n$.

(iii) If the greatest E is required, for example where the dielectric stress may cause breakdown, some improvement may result by cell subdivision, particularly where change in cell size is rapid. This cannot be carried too far, however, for the accuracy of measuring l and b decreases as the cell is made smaller.

8.5   Numerical methods

These can be used for regions with boundaries of any shape, particularly where analytical methods fail. They may be adapted for digital computers and thus produce solutions with any degree of accuracy. Here we shall discuss only the finite difference methods of iteration and relaxation as applied to two-dimensional Laplacian regions.

8.5.1   Finite-difference solution

Figure 8.6a shows a grid of uniform size h. We consider the voltage gradients at the midpoints A, B, C, D. Clearly

$$\left.\frac{\Delta V}{\Delta x}\right|_A = \frac{V_1 - V_0}{h} \ ; \quad \left.\frac{\Delta V}{\Delta x}\right|_C = \frac{V_0 - V_3}{h}$$

and from these

$$\left.\frac{\partial^2 V}{\partial x^2}\right|_0 = \left(\left.\frac{\Delta V}{\Delta x}\right|_A - \left.\frac{\Delta V}{\Delta x}\right|_C\right)\Big/h = \frac{V_1 + V_3 - 2V_0}{h^2}$$

Likewise

$$\left.\frac{\partial^2 V}{\partial y^2}\right|_0 = \left(\left.\frac{\Delta V}{\Delta y}\right|_A - \left.\frac{\Delta V}{\Delta y}\right|_D\right)\Big/h = \frac{V_2 + V_4 - 2V_0}{h^2}$$

Hence

$$\frac{\partial^2 V}{\partial x^2} + \frac{\partial^2 V}{\partial y^2} = \frac{V_1 + V_2 + V_3 + V_4 - 4V_0}{h^2} = 0$$

Thus the finite difference criteria for satisfying Laplace's equation is

$$V_0 = \tfrac{1}{4}(V_1 + V_2 + V_3 + V_4)$$

Figure 8.6b shows an uneven mesh that may occur at boundaries and corners. Proceeding in the same way, but not with the same ease, it can be shown that the

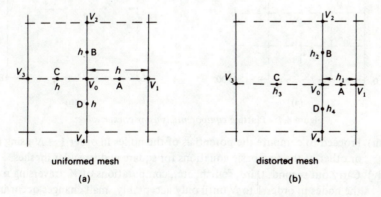

uniformed mesh   distorted mesh

(a)   (b)

Figure 8.6   *Finite-difference solution grid.*

finite difference critieria can be written

$$V_0 = \cfrac{V_1}{\left(1 + \cfrac{h_1}{h_3}\right)\left(1 + \cfrac{h_1 h_3}{h_2 h_4}\right)} + \cfrac{V_2}{\left(1 + \cfrac{h_2}{h_4}\right)\left(1 + \cfrac{h_2 h_4}{h_1 h_3}\right)}$$

$$+ \cfrac{V_3}{\left(1 + \cfrac{h_3}{h_1}\right)\left(1 + \cfrac{h_3 h_1}{h_2 h_4}\right)} + \cfrac{V_4}{\left(1 + \cfrac{h_4}{h_2}\right)\left(1 + \cfrac{h_4 h_2}{h_3 h_1}\right)}$$

Note the subscript symmetry in the denominators. By writing $h = h_1 = h_2 = h_3 = h_4$, this condition is the same as the former.

The iteration and relaxation processes apply these equations systematically and repeatedly to improve the computed values of potential V at the nodes of a mesh in the region. Always identify V_0 with the required potential.

8.5.2 Iteration process

This can be summarised as follows

(i) Mark out the field region in a convenient square-meshed grid and number the nodes 1 to N.

(ii) Obtain initial values of potential for the N nodes by guessing, estimating, using a free-hand sketch, etc. It is sometimes possible to apply the finite-difference equations diagonally on a larger cell size. For instance, as shown in figure 8.7a we would normally compute V_0 from V_1, V_2, V_3, V_4; but these being unknown initially we can take $V_0 = \frac{1}{4}(0 + 0 + 100 + 50) = 37\cdot5$.

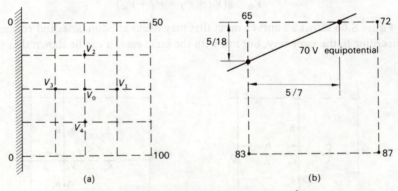

Figure 8.7 *Plotting equipotentials using corner values.*

(iii) Proceed to compute the potentials of the nodes in order $1 \rightarrow N$ using one or other of the two basic equations for square and irregular meshes.

(iv) Carry out second, third, fourth, etc., computations of V, traversing round the nodes in order 1 to N until only acceptably small changes occur in successive computations. The process is slowly convergent.

(v) If a map of equipotentials is required, locate the positions of a required potential by proportionality, as shown in figure 8.7b where the 70 V equipotential line is being located and mapped.

(vi) If the orthogonal flow-line map is also needed, this can be obtained from the potential map by the method of curvilinear squares; for instance by making $\Delta l = \Delta b$ by successive adjustments. An alternative would be to set up the dual problem by interchanging equipotential (source, sink) boundaries with the insulating boundaries and to carry out on it another iteration process. This would effectively compute values of u for the given grid.

8.5.3 Iteration example

The particular example is shown in figure 8.8 and consists of an earthed cylinder of internal diameter 8 cm and a central conductor 2 cm by 2 cm section. A quarter-section is shown, though strictly only an eighth is necessary, the remainder having symmetrical values. There are only 6 unique or basic nodes in a 1 cm grid; of these nodes 1, 4, 5, 6 are of a uniform grid. Nodes 2 and 3 are of uneven mesh, requiring the use of the second V_0 equation; the multiplying factors are given. In the figure the original assigned values are given in the first quadrant. Second, third and fourth computer values are given in the second, third and fourth quadrants. Note that the basic nodes were traversed in sequence order 1, 2, 3, 4, 5, 6; the

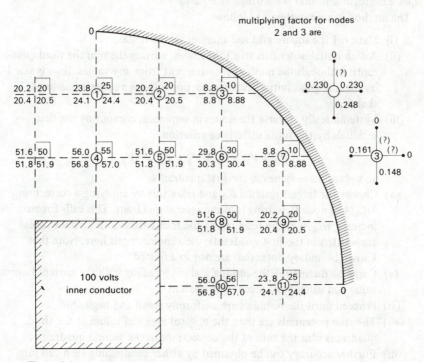

Figure 8.8   *Iteration example: nodes 7, 8, 9, 10, 11, repeat 3, 5, 2, 4, 1.*

second traverse started again at node 1. Other nodes are repetitions of the basic nodes.

In computing a new value of V_0 the most up-to-date values of V_1, V_2, V_3 and V_4 are always used.

For a 0·5 cm mesh, the intermediate potentials could be found straight away from the final values at the main nodes.

A computer program is given in Appendix 4.

8.5.4   Relaxation process

This systematic approach to the solution of a set of linear equations was due to Hardy-Cross (1932) and Southwell (1935) and many other workers were also involved. It arose originally in calculating stresses in redundant pin-jointed frameworks; but it has also been extensively employed in many physical problems including the solution of electrical networks.

As regards the two-dimensional laplacian field, the method is like iteration in that it finds successively better approximations for the potentials at the nodes in the grid. However, it operates by dissipating the residual errors rather than adjusting the full potential values. Although the method can be used for irregular field patterns we shall consider it here only for regions based entirely on the square uniform cell, that is no irregular boundaries.

The method can be outlined as follows.

(i) Mark off the square grid and number all the nodes.

(ii) Assign initial potentials to all the nodes, writing them in the third quadrants. Although the method is convergent from any values, time is saved by making good initial assignments, possibly even using a free-hand sketch.

(iii) Systematically traverse the nodes in sequence, computing the first residuals by the finite difference criterion

$$R_0 = V_1 + V_2 + V_3 + V_4 - 4V_0$$

and place these values in the first quadrants.

(iv) Choose the largest residual R_0 and relax this by applying a correction $+R_0/4$ at node 0, writing it in the second quadrant. This calls for an increase $+R_0/4$ at the residuals of the four neighbouring nodes. These are written in the first quadrants; use running totals here. Note that known boundary potentials are not so adjusted.

(v) Carry on dissipating the residuals, always dealing with the current largest value and using most recent information.

(vi) Proceed until the residuals are uniformly small and negligible.

(vii) The true potentials are then the original assigned values in the third quadrants plus the sum of the corrections in the second quadrants.

(viii) Further accuracy can be obtained by either subdividing each cell into four and starting again using the first solution or alternatively increasing

the potentials and residuals by a factor of 10 and further dissipating
the residuals.

(ix) One word of warning! No errors should be made in the arithmetic for,
unlike iteration, any such error is not automatically self-cancelling later.

Part of a grid is shown in figure 8.9. Suppose the initial assigned values are as
shown in the third quadrants. Consider node 5; its residual is

$$R_5 = 54 + 78 + 80 + 50 - 4 \times 60 = 22$$

and suppose this as well as the other residuals are all as recorded in the first
quadrants. Choose $R_5/4 = +5$ to be the correction to apply to V_5 and enter this

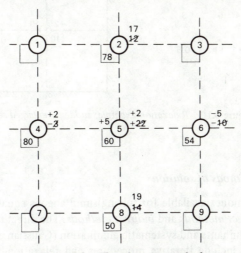

Figure 8.9   *The relaxation method.*

in the second quadrant. Now adjust R_5 by subtracting 20, and R_6, R_2, R_4, R_8
by adding 5 to each in their first quadrants. Node 8 would next be dealt with,
and so on.

8.5.5   *Relaxation example–suddenly enlarged section*

The example of the relaxation process in figure 8.10 is of a suddenly enlarged
section. The potential values given at the open ends of the section are held constant. The assigned values are shown in the third quadrants, and the initial residues
in the first quadrants. The diagram indicates the first 9 relaxations, giving the node
and correction values. A fairly accurate result is obtained after about 40 relaxations;
see table 8.2.

Table 8.2

Node	1	2	3	4	5	6	7	8	9
Potential	50·24	50·95	53·57	64·32	70·19	29·53	42·42	12·37	19·95
Residual	−0·01	+0·01	−0·01	+0·01	−0·02	−0·01	−0·01	0	−0·01

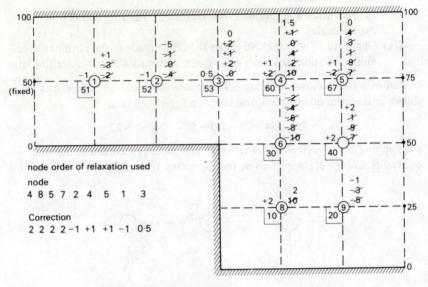

Figure 8.10   *Relaxation example: suddenly enlarged section.*

8.5.6   *Direct methods of solution*

The various techniques available for solving simultaneous equations may be classified into *direct methods* and *indirect methods*. The direct methods include solution by determinants and systematic elimination (Gaussian elimination). The indirect methods include iterative procedures and relaxation techniques, as discussed previously. Here we discuss the direct approach, and we do this by way of an example.

8.5.6.1   *Poissonian field*   Consider the two-dimensional Poissonian field shown in figure 8.11. The field is described by

$$\nabla^2 V = -\frac{\rho_V}{\epsilon} = 80$$

In terms of the finite-difference equation this becomes

$$\frac{\Delta^2 V}{\Delta x^2} + \frac{\Delta^2 V}{\Delta y^2} = \frac{1}{h^2} \; [V_1 + V_2 + V_3 + V_4 + 4V_0] = 80$$

In this example there are only three unique values of V. Nodes 2, 4, 6 and 8 are equivalent and relabelled node 2. Nodes 1, 3, 7 and 9 are also equivalent and relabelled node 3. The potentials at nodes 1 to 9 are therefore obtained by finding the potentials at the equivalent nodes 1, 2 and 3. Three equations are therefore necessary and these are found from the above finite-difference formula

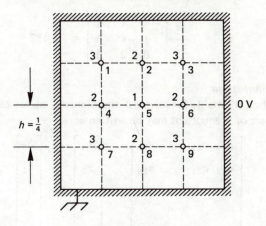

Figure 8.11 *Direct solution example of Poissonian field where* $\nabla^2 V = -\rho_V/\epsilon = 80$.

node	equation
1	$-4V_1 + 4V_2 + 0V_3 = h^2 \times 80 = 5$
2	$+V_1 - 4V_2 + 2V_3 = 5$
3	$0V_1 + 2V_2 - 4V_3 = 5$

(i) *Solution by determinants*

This is the simplest approach for this problem. However, for many equations this approach is not suitable for use on a computer because of the large number of multiplications necessary. In the above problem the determinant of the coefficients is

$$\left| D_c \right| = \begin{vmatrix} -4 & 4 & 0 \\ +1 & -4 & 2 \\ 0 & 2 & -4 \end{vmatrix} = -32$$

The potentials are then

$$V_1 = \frac{5}{-32} \begin{vmatrix} 1 & 4 & 0 \\ 1 & -4 & 2 \\ 1 & 2 & -4 \end{vmatrix} = -5.625$$

$$V_2 = \frac{-5}{32} \begin{vmatrix} -4 & 1 & 0 \\ 1 & 1 & 2 \\ 0 & 1 & -4 \end{vmatrix} = -4.375$$

$$V_3 = \frac{-5}{32} \begin{vmatrix} -4 & 4 & 1 \\ 1 & -4 & 1 \\ 0 & 2 & 1 \end{vmatrix} = -3.437$$

(ii) *Gaussian elimination*

This direct method is suitable for solving N simultaneous equations on a small computer. The set of N equations may be written as

$$\begin{bmatrix} a_{11} & a_{12} & \cdots & a_{1n} \\ a_{21} & a_{22} & \cdots & a_{2n} \\ \vdots & \vdots & & \vdots \\ a_{n1} & a_{n2} & & a_{nn} \end{bmatrix} \cdot \begin{bmatrix} x_1 \\ x_2 \\ \vdots \\ x_n \end{bmatrix} = \begin{bmatrix} b_1 \\ b_2 \\ \vdots \\ b_n \end{bmatrix}$$

which may be abbreviated to

$$A \cdot x = b$$

This involves solving the set of equations

$$x = A^{-1} \cdot b$$

where the a_{ij} and b_i are given and the x_i are to be found.

The systematic elimination of variables is then achieved by dividing the first equation ($i = 1$) in the set by a_{11}, followed by multiplication of this equation by a_{21}. The first equation is then substracted from the second to generate a new equation, with new coefficients but with the first variable eliminated. This process is repeated on the remaining equations and finally a new set of equations is obtained of the form

$$x_1 + c_{12}x_2 + \cdots \cdots \cdots \cdots c_{1n}x_n = d_1$$
$$x_2 + \cdots \cdots \cdots \cdots c_{2n}x_n = d_2$$
$$\vdots \qquad\qquad \vdots \qquad \vdots$$
$$x_n = d_n$$

The final equation gives the solution for x_n and the remaining variables are determined by back substitution. Although the technique is relatively simple when solving two or three equations by hand, the programming necessary for digital computation of a large set of equations has to consider logical problems which the individual solves intuitively. A simple program written in BASIC, which solves N simultaneous equations, is given in Appendix 4. Running this program for the above problem gives $V_1 = -5.625$, $V_2 = -4.375$ and $V_3 = -3.4375$, which is in agreement with the solution obtained using determinants.

Problems

8.1 A long metal trough has inside dimensions 6 in by 9 in; centrally inside is a
1000 V d.c. cylindrical conductor of 3 in diameter. *Sketch* the field over one
quarter section.
(a) Given $\epsilon_r = 6$ for the dielectric, estimate capacitance μF/km.
(b) Find the greatest values of E at trough and conductor surfaces and the
corresponding values of surface charge density ρ_S.
(c) If the dielectric resistivity is $2 \times 10^8 \ \Omega$ m, find the insulation resistance,
Ω/km.

((a) 0.325μ F/km; (b) 2.23×10^4 V/m; 3.04×10^4 V/m; 1.6μC/m^2,
1.18μC/m^2; (c) $32.6 \times 10^3 \ \Omega$/km)

8.2 The sketch, figure P8.2, shows the half cross-section of a very long conducting
busbar and an earth plane; dimensions are in inches. The bar has 4000 V
potential above earth. Sketch its field.

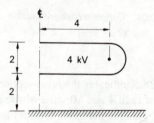

Figure P8.2 *Busbar and earth plane.*

(a) Find its capacitance in F/m.
(b) Compare it with a plate 10 inches wide, 2 inches from the plane, ignoring
fringing.
((a) $7.1\epsilon_0$ F/m; (b) $5\epsilon_0$ F/m)

8.3 Use an iterative process to compute the potentials of the 11 distinct points
in the cm grid of the two-dimensional problem indicated in figure P8.3; only
one-quarter is shown.

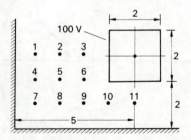

Figure P8.3 *Two-dimensional field problem, iterative process.*

($14.75, 35.46, 60.67, 12.80, 29.26, 54.62, 7.24, 16.29, 28.69, 43.92, 46.96$)

8.4 The diagram figure P8.4 shows a 2 by 1 corner with 1 cm grid. Use a relaxation computation to find the potentials at points A, B, C along the corner diagonal.

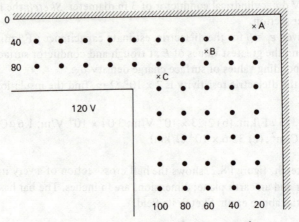

Figure P8.4 *Corner section, relaxation example.*

(2·6, 23·3, 70·6)

8.5 The diagram figure P8.5 indicates the cross-section of two long, conducting cylindrical surfaces of radii 4 and 10 units whose axes, though parallel, are displaced by 2 units. Assume the inner cylinder is held at 4 kV and the outer is earthed. Make a curvilinear field sketch showing equipotentials at 1, 2 and 3 kV. Use the sketch to estimate

(a) the capacitance per mile (1600 m) assuming $\epsilon_r = 3$ and
(b) the insulation resistance per mile assuming resistivity $\rho = 10^{12}$ Ω m.

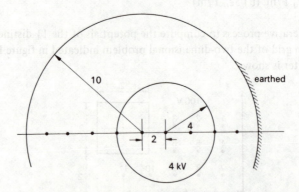

Figure P8.5 *Two long, non-coaxial, conducting cylinders.*

(0·31μ F/mile, 85 MΩ/mile)

9

Field Problems—Analytical Solutions

The previous chapter has dealt with the main 'non-exact' methods of solving laplacian field problems, particularly of two dimensions. The application of numerical analysis to field problems and the advent of the digital computer have enabled complex field problems to be solved. However, these rely heavily on computer facilities, programming experience and particularly on the suitability of the model used. Here one should note that it is not always desirable to model fields in terms of circuits and networks since this may well lead to a loss of physical insight into the nature of a particular field problem.

We now consider some of the analytical or 'exact' methods available for solving such problems.

9.1 Kelvin's method of images

This was introduced by Kelvin, in 1848, for solving electrostatic problems in which conducting boundary surfaces were present. Such a surface is shown in figure 9.1 in which we have region I containing two point charges Q_1, Q_2 and a

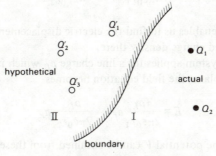

Figure 9.1 *Kelvin's method of images.*

173

conducting equipotential boundary surface separating it from region II. The potential of the boundary is often chosen to be zero. It is possible to find a set of charges in region II, $Q_1', Q_2', Q_3', \ldots$, so disposed that when the physical boundary is removed there is an exactly similar equipotential V formed by the charge system, and the field in region I remains unaltered.

The utility of the method lies entirely in the simplicity of the boundary surface and the ease with which the values and positions of the hypothetical charges $Q_1', Q_2', \ldots$ can be located.

There are some points of similarity with the images of light sources in mirrors but the duality often breaks down, particularly when iron boundaries and positive and negative currents are considered. A few examples will be discussed to illustrate the method.

9.1.1 Point charge and conducting plane

This is the classic problem and a point charge in the vicinity of a conducting plane is shown in figure 9.2a.

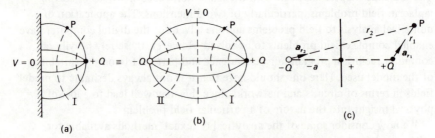

Figure 9.2   *Point charge and conducting plane.*

The conducting plane and its induced electric surface charge produce the same field in region I as a charge $-Q$ placed in the mirror-image position and the boundary plane removed. The field is illustrated in figure 9.2b and its strength at any point P (figure 9.2c) is given by

$$E = \frac{Q}{4\pi\epsilon_0 r_1^2} a_{r_1} - \frac{Q}{4\pi\epsilon_0 r_2^2} a_{r_2}$$

In particular this enables us to find the electric displacement D at the boundary plane and the induced charge density there.

The same image system applies for a line charge ρ_L which is parallel to an infinite conducting plane; the field equation becomes

$$E = \frac{+\rho_L}{2\pi\epsilon_0 r_1} a_{r_1} - \frac{\rho_L}{2\pi\epsilon_0 r_2} a_{r_2}$$

Expressions for the potential V can be obtained from these equations since $E = -\,\mathrm{grad}\,V$.

9.1.2   Point charge and sphere

The point charge Q_1 is shown in figure 9.3 at a distance c from the centre of the sphere of radius b. This is not a very realistic problem but it brings out principles. Consider two point charges Q_1 and Q_2 with $Q_1 > Q_2$. The potential at any point P is

$$V = \frac{1}{4\pi\epsilon_0}\left(\frac{Q_1}{r_1} + \frac{Q_2}{r_2}\right)$$

Consider an equipotential surface of $V = 0$ through P, then

$$4\pi\epsilon_0 V = 0 = \frac{Q_1}{r_1} + \frac{Q_2}{r_2}$$

For the surface, $r_2/r_1 = -Q_2/Q_1 = \alpha$ say. The section shown is the 'circle of Apollonius'. The surface is spherical about Q_2 but is *not* centred on Q_2. Consider points P′ and P″

$$\alpha = \left(\frac{r_2}{r_1}\right)' = \frac{b-d}{c-b} \quad \text{and} \quad \alpha = \left(\frac{r_2}{r_1}\right)'' = \frac{b+d}{c+b}$$

By first equating these and then multiplying them, we obtain displacement of charge from centre $d = b^2/c$, ratio $\alpha = b/c = -Q_2/Q_1$, so that $Q_2 = -(b/c)Q_1$, distance $a = c - d = c - b^2/c = b/\alpha - b\alpha = b[(1 - \alpha^2)/\alpha]$. Hence, if we know Q_1, the radius

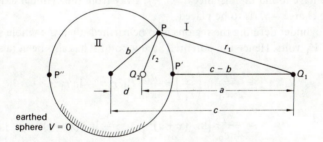

Figure 9.3   *Image system for point charge and sphere.*

b and the distance away c of the sphere, we can compute the image point charge $Q_2 = -(b/c)Q_1$ and its displacement $d = b^2/c$ from the centre of the sphere.

We are now in a position to compute the potential V and the field strength E at any point in region I due to a point charge Q_1 outside a conducting sphere.

The problem is applicable to the discharge at sphere gaps.

9.1.3   Line charge and conducting cylinder

This problem has already been solved in section 3.6.3 and some of the previous results are used below. We refer to figure 9.4 in which we have a line charge ρ_L, distant f, from the centre of a cylinder of radius b.

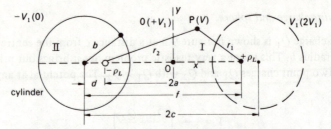

Figure 9.4 *Image system for line charge and cylinder.*

We wish to find where to put the image charge $-\rho_L$ within the sphere to give an unchanged field in region I. Consider the line charge $-\rho_L$ located at $2a$ away and the two equipotential cylindrical surfaces $\pm V_1$.

$$\frac{r_2}{r_1} = (K_1)^{1/2} = \exp\left(\frac{2\pi\epsilon_0 V_1}{\rho_L}\right)$$

$$= \frac{b}{c-a} = \frac{c+a}{b} = \frac{b}{d} = \frac{f}{b}$$

for

$$\left.\begin{array}{ll} 2a = f - d & c + a = f \\ 2c = f + d & c - a = d \end{array}\right\} \quad c^2 - a^2 = b^2 = fd$$

Hence we have found the distance $d = b^2/f$, away from the cylinder axis, at which the image charge $-\rho_L$ is to be placed.

If the cylinder defining the regions is to be earthed, then the whole system is raised by V_1 volts. Hence the potential at any point P has also been raised by V_1 and now

$$V - V_1 = \frac{\rho_L}{2\pi\epsilon_0} \ln\left(\frac{r_2}{r_1}\right) = \frac{\rho_L}{4\pi\epsilon_0} (\ln r_2{}^2 - \ln r_1{}^2)$$

$$= \frac{\rho_L}{4\pi\epsilon_0} [\ln \{(x+a)^2 + y^2\} - \ln \{(x-a)^2 + y^2\}]$$

From this

$$E_x = -\frac{\partial V}{\partial x} = \frac{-\rho_L}{4\pi\epsilon_0}\left(\frac{1}{r_2{}^2}\frac{\partial r_2{}^2}{\partial x} - \frac{1}{r_1{}^2}\frac{\partial r_1{}^2}{\partial x}\right)$$

$$= \frac{\rho_L}{2\pi\epsilon_0}\left(\frac{x-a}{r_1{}^2} - \frac{x+a}{r_2{}^2}\right)$$

$$E_y = -\frac{\partial V}{\partial y} = \frac{-\rho_L}{4\pi\epsilon_0}\left(\frac{1}{r_2{}^2}\frac{\partial r_2{}^2}{\partial y} - \frac{1}{r_1{}^2}\frac{\partial r_1{}^2}{\partial y}\right)$$

$$= \frac{\rho_L}{2\pi\epsilon_0}\left(\frac{1}{r_1{}^2} - \frac{1}{r_2{}^2}\right)$$

Thus the whole field pattern can be computed.

9.1.4   Line or point charge and two intersecting planes

This case has some similarity to that of light images with two plane mirrors; multiple images are produced. Two intersecting planes are shown in figure 9.5 in which the angle between the planes is θ. Three cases are illustrated, one for θ/π, non-integer, and two for θ/π, integer.

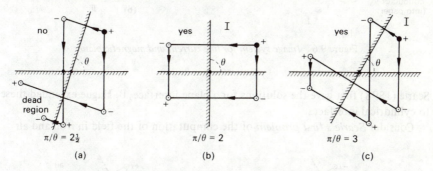

Figure 9.5   *Image systems for charge and two intersecting planes.*

To replace the effect of two conducting planes carrying induced charges due to a point or line charge between them is only possible when π/θ *is an integer*; otherwise the two trains of reflections terminate in two non-coincident charges in the 'dead region' where reflections must cease.

When π/θ is an integer only one charge is produced in the dead region and the image system correctly replaces the two planes and their charges as regards region I.

Clearly the initial charge and the images all lie on the circumference of a circle. In the diagrams shown, the first figure 9.5a has $\pi/\theta = 2\frac{1}{2}$ and produces an image system which is not acceptable. The other two figures 9.5b and c, for which π/θ is 2 and 3, produce acceptable image systems from which the region I potential and field strength can be computed.

9.1.5   Line current and magnetic plane

Electrical machines have air spaces and iron regions with conductors in air and embedded in the iron. Two main problems arise: these are the computation of the magnetic field arising in the air and in the iron due to a conductor in the iron and in the air. The image system is shown in figure 9.6. The two diagrams show the image conditions required for computing the field in regions I (where the actual conductor is) on the assumption that $\mu_r \to \infty$.

For intersecting magnetic planes, the boundary conditions can be satisfied only if π/θ is an integer. Again two cases arise, which are indicated in figure 9.7 for $\pi/\theta = 3$; these assume $\mu_r \to \infty$ and only fields in regions I are needed. The realistic problems, however, are those arising when μ_r is fairly large but not infinite. G. F. C.

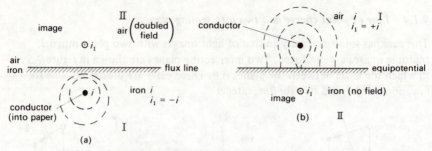

Figure 9.6   *Image system for line current and magnetic plane.*

Searle (1898) first gave the solutions for a plane interface; B. Hague extended these to cylindrical interfaces.

Consider *Searle's first problem* of the computation of the field in iron and air

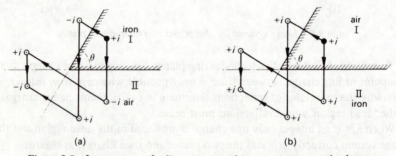

Figure 9.7   *Image system for line current and intersecting magnetic planes.*

when one conductor is embedded in iron of constant permeability $\mu \neq \infty$. This is shown in figure 9.8a. For ease, we use $\mu = \mu_r$ here.

Searle set about the problem in an interesting way. For the field in the iron he assumed all space to be iron and an image current i_1 at B. For the field in the air he assumed all space to be air and only one modified current i_2 (larger than i) to be at A, with nothing at B.

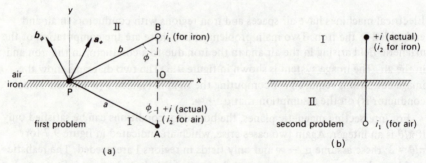

Figure 9.8   *Searle's two problems.*

To compute i_1 and i_2 consider any point P at the interface. The three currents i, i_1 and i_2, have fields at P which will be $(i/2\pi a)\mathbf{a}_\phi$, $(i_1/2\pi b)\mathbf{b}_\phi$ and $(i_2/2\pi a)\mathbf{a}_\phi$ where $|a| = |b|$. These can be resolved horizontally and normally giving

horizontally $\dfrac{i}{2\pi a}\cos\phi,$ $-\dfrac{i_1}{2\pi a}\cos\phi,$ $\dfrac{i_2}{2\pi a}\cos\phi$

normally $\dfrac{i}{2\pi a}\sin\phi,$ $\dfrac{i_1}{2\pi a}\sin\phi,$ $\dfrac{i_2}{2\pi a}\sin\phi$

In the actual situation at point P at the interface, the tangential component of the magnetic field strength H_t is the same in both media and the normal component of magnetic flux density B_n is continuous. Hence the conditions are

$$H_t:\qquad i - i_1\quad = i_2$$

$$B_n:\qquad \mu(i + i_1) = i_2$$

Solving these two equations for i_1 and i_2

$$i_1 = \frac{-i(\mu - 1)}{\mu + 1} \qquad\text{and}\qquad i_2 = \frac{i(2\mu)}{\mu + 1}$$

To compute the field in the iron we use i and i_1; for the field in the air we only use i_2; note that, when $\mu \to \infty$, $i_1 = -i$ and $i_2 = 2i$; the field in air is doubled.

For *Searle's second case*, when the current i is in air, an exactly similar process is followed (see figure 9.8b). The corresponding currents are $i_1 = + i(\mu - 1)/(\mu + 1)$ and $i_2 = i(2)/(\mu + 1)$. To compute the field in the air we use i and i_1; for the field in the iron we use only i_2, which in this case is smaller than i. Note that, if $\mu \to \infty$, $i_1 = +i$ and $i_2 \to 0$ so that there is no field in the iron.

Some little care is needed in applying these results.

9.1.6   *Example–conductor between two magnetic plates*

Consider one linear current i between two parallel magnetic faces as shown in figure 9.9. The spacing between the plates is $3a$ with the current element a distance a from one plate. Let $\mu_r \to \infty$. An infinite series of image currents, all $+i$ (into paper) is produced. The field at the original current due to all the images will

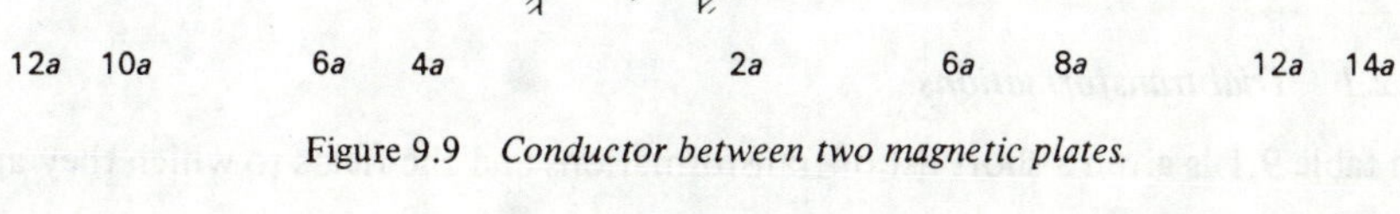

Figure 9.9   *Conductor between two magnetic plates.*

be in the y direction and the magnetic field strength is given by

$$H_y = \frac{i}{2\pi a} \left[\left(\frac{1}{2} - \frac{1}{4}\right) + \left(\frac{1}{6} - \frac{1}{6}\right) + \left(\frac{1}{8} - \frac{1}{10}\right) + \left(\frac{1}{12} - \frac{1}{12}\right) + \left(\frac{1}{14} - \frac{1}{16}\right) + \cdots \right] \text{A/m}$$

$$= \frac{i}{4\pi a} \left[\left(1 - \frac{1}{2}\right) + \left(\frac{1}{4} - \frac{1}{5}\right) + \left(\frac{1}{7} - \frac{1}{8}\right) + \left(\frac{1}{10} - \frac{1}{11}\right) + \cdots \right]$$

$$= \frac{i}{4\pi a} \left[\frac{1}{1 \times 2} + \frac{1}{4 \times 5} + \frac{1}{7 \times 8} + \frac{1}{10 \times 11} + \cdots \right]$$

$$= \frac{1}{4\pi a} (0 \cdot 6001) \quad \text{(to 30 terms)}$$

For the force F, $\mathrm{d}F = i(\mathrm{d}l \times B)$; $\mathrm{d}l = \mathrm{d}l(-a_z)$. $B = \mu_0 H_y a_y$ from which

$$F = F_x a_x \frac{\mu_0 i^2}{4\pi a} (0 \cdot 6001) \text{ N/m}$$

Hague gives an exact analytical solution to this problem.

9.2 Conformal transformation

We have $z = x + jy$ as defining a point in a two-dimensional field. The related field function $w = u + jv = f(z)$ defines the flux function u and the potential function v at the point. We know that u and v are conjugate functions each satisfying Laplace's equation in two dimensions. The main problem is that of finding the transforming function $w = f(z)$ to fit the known particular boundary requirements.

Two main methods exists for determining w: (i) by trial transformations and (ii) by Schwarz–Christoffel method.

The first of these is really a trial and error process; one systematic method is to choose w, map u and v and then decide what boundary-value problem it solves. A list of such transforms can be made and used for reference.

The *Schwarz–Christoffel* transformation starts from any boundary system which is *a linear polygon* in the z plane; it transforms it systematically to a t plane, as intermediate, before finally fitting it to a w plane. We shall not develop this here.

9.2.1 Trial transformations

In table 9.1 is given a short list of transformations and the fields to which they apply.

Table 9.1

Transform	Field
$w = z^{-1}$	electric or magnetic dipole
$z + z^{-1}$	flow round a hole in a plate
$\ln (z - a)/(z + a)$	field due to a pair of parallel cylinders
$\ln z$	field of a line charge
z^2	flow inside a $90°$ bend
$z^{1/2}$	field of a charged semi-infinite plate
$\sin z$	pole shape for sine flux density
$\ln \sin z$	set of parallel wires all similarly charged
$\cosh^{-1} z$	flow through a slit
$\exp z$	temperature distribution in a plate
z^{-2}	quadripole line; charged in sequence $(+-+-)$
$\cosh z$	confocal ellipses and hyperboles
$\sin^{-1} z$	gap in a charged infinite plate
$z^{2/3}$	$90°$ magnetised corner
$z = w + \exp w$	edge effect in capacitor plates
$z = a \cos w + jb \sin w$	charged elliptical cylinder
$w = (az + b)/cz + d)$	bilinear transform for various field operations
$\ln \sin z - \ln \cos z$	parallel wires charged alternately
$\cos^{-1} z$	as $\sin^{-1} z$ but different orientation

9.2.2 $w = u + jv = z^{-1}$

Consider $w = u + jv = z^{-1}$ and find out the field problem to which it applies. Separate out the real and imaginary terms of $1/z$. Do not confuse the complex variable $z(= x + jy)$ with the cartesian coordinate z.

$$z^{-1} = \frac{1}{z} = \frac{1}{x + jy} = \frac{x - jy}{x^2 + y^2} = \frac{x}{x^2 + y^2} - \frac{j(y)}{x^2 + y^2}$$

$$= u - jv$$

$$u = \frac{x}{x^2 + y^2}$$

$$x^2 + y^2 - \frac{x}{u} = 0 \qquad \text{(circles, centre } (1/2u, 0); \text{ diam. } 1/u)$$

$$v = \frac{-y}{x^2 + y^2}$$

$$x^2 + y^2 + \frac{y}{v} = 0 \qquad \text{(circles, centre } (0, -1/2v); \text{ diam. } 1/v)$$

Figure 9.10a,b shows the true field in the z plane drawn for $u = 0, \pm 0.5, \pm 1$ and $v = 0, \pm 0.5, \pm 1$ and the transform to a square grid in the w plane. The curvilinear square from P in the z plane corresponds exactly to the true square from Q in

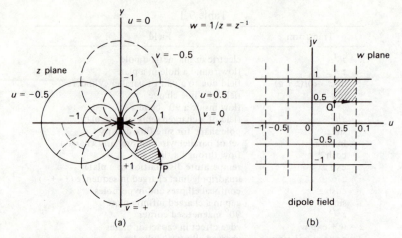

Figure 9.10   *True field in xy plane and transform to square grid in w plane.*

the w plane. The transformation relates to an electric (or magnetic) dipole on the y axis; u is the flux function Ψ and has the direction of the electric field strength E and v is the potential V.

9.2.3 $w = z^{2/3}$

This is another instructive transform, $w^3 = z^2$. We can proceed as before.

$$w = u + jv = z^{2/3} = (r \exp j\theta)^{2/3} = (r^{2/3}) \exp j2\theta/3 = r^{2/3}(\cos \theta_w + j \sin \theta_w)$$

Here $\theta_w = 2\theta/3$ has been introduced. Separate variables

$$u = r^{2/3} \cos \theta_w; \qquad r = u^{3/2}(\sec \theta_w)^{3/2}$$

$$v = r^{2/3} \sin \theta_w; \qquad r = v^{3/2}(\operatorname{cosec} \theta_w)^{3/2}$$

To plot the fields from these take values $v = 0, 0 \cdot 5, 1, 1 \cdot 5$ and $u = 0, \pm 0 \cdot 5, \pm 1, \pm 1 \cdot 5$. Assume that for the negative values of u, r has only positive real values. Table 9.2 shows a few restricted values for θ, u, v; figure 9.11 shows the sketched field in the z plane. The w plane is as before (figure 9.10b).

Table 9.2

$\theta(= 1 \cdot 5\theta_w)$	0	$45°$	$90°$	$135°$	$180°$	$225°$	$270°$	
$(\sec \theta_w)^{1 \cdot 5}$	1	$1 \cdot 242$	$2 \cdot 828$	$\pm \infty$	$-2 \cdot 828$	$-1 \cdot 242$	$-1 \cdot 0$	
$(\operatorname{cosec} \theta_w)^{1 \cdot 5}$	∞	$2 \cdot 828$	$1 \cdot 242$	1	$1 \cdot 242$	$2 \cdot 828$	$\pm \infty$	
r	∞	$5 \cdot 20$	$2 \cdot 28$	$1 \cdot 84$	$2 \cdot 28$	$5 \cdot 20$	∞	$v = 1 \cdot 5$
r	$0 \cdot 354$	$0 \cdot 439$	$1 \cdot 00$	$\pm \infty$				$u = 0 \cdot 5$

The problem solved is that for an isolated, $90°$ magnetised corner; the origin of coordinates is at the corner. It could apply also to the flow from a right-angled source with the sink at infinity.

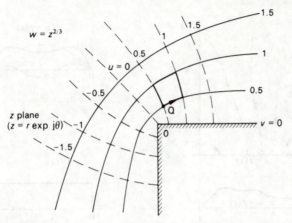

Figure 9.11   *Conformal transformation $w = z^{2/3}$; magnetised 90° corner.*

9.3   Separation of variables

All solutions to Laplace's equation, $\nabla^2 V = 0$, in a given region are known as harmonic functions; the problem is to find the harmonic function which satisfies the boundary conditions. Any solution to the potential equation must be a harmonic function or a finite or infinite series of such functions. These functions can be expanded into convergent power series and they are therefore analytic.

The particular harmonic function employed will depend on the type of symmetry and the coordinate system employed. In these cases which are soluble by analysis, the method of separation of variables is used. This will be shown by consideration of three typical cases

    (i) cartesian coordinates—Fourier series.
    (ii) cylindrical coordinates—circular harmonics.
    (iii) spherical coordinates—Legendre polynomials.

9.3.1   *Cartesian coordinates—Fourier series*

Consider the current flow in a parallel-sided semi-infinite plate to which a known potential (voltage) function

$$V_{y=0} = f(x)$$

is applied along the x edge. The semi-infinte plate is shown in figure 9.12a. As a solution try $V = XY$ in which $X = f_1(x)$ and $Y = g_1(y)$, that is each is a function of one coordinate only.

$$\nabla^2 V = \frac{\partial^2 V}{\partial x^2} + \frac{\partial^2 V}{\partial y^2} = X''Y + XY'' = 0 \qquad \text{(Laplace)}$$

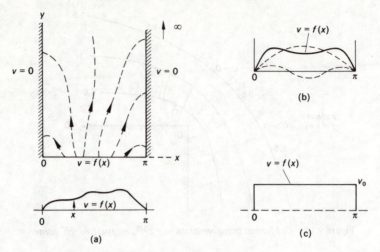

Figure 9.12   *Current flow in a long plate.*

Hence $X''/X = -Y''/Y = -n^2$ (say) where n = constant. This is a subtle deduction for $X''/X = F(x) + a$ and $Y''/Y = G(y) + b$ and therefore, since a function of x cannot equal a function of y, we must have $F(x) = 0$ and $G(y) = 0$, giving $-a = b = n^2$. The variables are separated and we have two second-order differential equations to solve

$$X'' + n^2 X = 0 \qquad \text{and} \qquad Y'' - n^2 Y = 0$$

The most general form of solution to these is

$$V = \sum_n (A_n \cos nx + B_n \sin nx)(C_n \exp(ny) + D_n \exp(-ny))$$

Now bring in the known boundary conditions. $V = 0$, $x = 0$, hence $A_n = 0$; $V = 0$, $x = \pi$, hence n = integer; $V \neq \infty$, $y \to \infty$, hence $C_n = 0$. This reduces the solution to

$$V = \sum_n a_n \exp(-ny) \sin(nx)$$

where $a_n = B_n D_n$. Whatever the known applied boundary function $V_{y=0} = f(x)$ may be, it has first to be expressed as a Fourier series; the coefficients a_n are then obtained by equating the coefficients of like powers.

Example 1   Let $V_{y=0} = \sin(x) + \frac{1}{4} \sin(3x)$ as shown in figure 9.12b. Then $a_0 = 0$, $a_1 = 1$, $a_2 = 0$, $a_3 = \frac{1}{4}$, giving the potential solution V at point (x, y)

$$V = \exp(-y) \sin(x) + \frac{1}{4} \exp(-3y) \sin(3x)$$

Example 2   Consider a rectangular wave as shown in figure 9.12c.

$$V_{y=0} = \frac{4V_0}{\pi} \left[\sin(x) + (\tfrac{1}{3}) \sin(3x) + (\tfrac{1}{5}) \sin(5x) + \cdots \right]$$

All the even coefficients are zero

$$a_0 = a_2 = a_4 \ldots = 0$$

All odd coefficients exist

$$a_1 = \frac{4V_0}{\pi}, a_3 = \frac{4V_0}{3\pi}, \ldots$$

This gives final solution

$$V = \frac{4V_0}{\pi} \sum_{n=1,3,\ldots} \left[\frac{1}{n} \exp(-ny) \sin nx \right]$$

Note that in computing the potential V for each point (x, y) in the plate there is an infinite series to evaluate. Probably the first three or four terms would be adequate.

The current function u can be determined by the use of conjugate functions as already explained.

9.3.2 Cylindrical coordinates–circular harmonics

Laplace's equation in cylindrical coordinates is

$$\nabla^2 V = \frac{1}{\rho} \frac{\partial}{\partial \rho} \left(\rho \frac{\partial V}{\partial \rho} \right) + \frac{1}{\rho^2} \frac{\partial^2 V}{\partial \phi^2} + \frac{\partial^2 V}{\partial z} = 0$$

If we are dealing only with problems independent of z, this becomes

$$\rho \frac{\partial}{\partial \rho} \left(\rho \frac{\partial V}{\partial \rho} \right) + \frac{\partial^2 V}{\partial \phi^2} = 0$$

As before let us assume a product-type solution $V = R(\rho)F(\phi)$ then

$$\rho F \frac{\partial}{\partial \rho} (\rho R') + RF'' = 0$$

Divide by RF

$$\frac{\rho^2 R'' + \rho R'}{R} + \frac{F''}{F} = 0$$

Arguing as before, both of these must be constants summing to zero. Using $\pm n^2$ as the constants we obtain

$$\rho^2 R'' + \rho R' - n^2 R = 0 \qquad \text{and} \qquad F'' + n^2 F = 0$$

Try a power series of ρ for the function R; a typical term is $R_n = A_n \rho^n$. Substituting

$$\rho^2 A_n n(n-1) \rho^{n-2} + \rho A_n n \rho^{n-1} - n^2 A_n \rho^n = 0$$

as required. Hence a power series in $\pm n$ will be suitable for R and

$$R = \sum_{n=1}^{\infty} (A_n \rho^n + B_n \rho^{-n})$$

The F equation is like the one obtained before giving

$$F = \sum_{n=1}^{\infty} [C_n \cos(n\phi) + D_n \sin(n\phi)]$$

The product $V = RF$ is *not* complete, for corresponding to $n = 0$ we can add any or all of the following terms

$$K_1 + K_2 \phi + K_3 \ln \rho + K_4 \phi \ln \rho$$

all of which satisfy the original Laplace equation. Finally

$$V = \sum_{n=1}^{\infty} (A_n \rho^n + B_n \rho^{-n})(C_n \cos \phi_n + D_n \sin \phi_n) + K_1 + K_2 \phi + K_3 \ln \rho + K_4 \phi \ln \rho$$

Boundary conditions will determine the constants K_1, K_2, etc.

Example—dielectric cylinder in extensive field   Suppose it is required to find the effect of placing a very long dielectric cylinder orthogonally to an extensive uniform electric field of strength E_0, see figure 9.13a. Suppose the original field strength E_0 to be in the x direction and take the $y = 0$ plane to be that of zero electric potential $V = 0$.

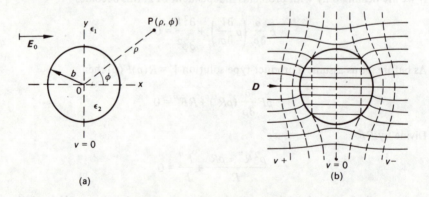

Figure 9.13   *Dielectric cylinder placed in extensive uniform electric field.*

Before the dielectric cylinder is put into position, from $E = -\text{grad } V$ the potential at point P is

$$V_{\text{p}} = -E_0 x = -E_0 \rho \cos \phi$$

When the cylinder *is* in position, the potential function will be the circular harmonic just discussed. From the symmetry of the problem, there can be no $\sin \phi$ terms present. Let us assume that for both regions 1, 2 the solution is of the form

$$V = \left(A\rho + \frac{B}{\rho}\right) \cos \phi$$

taking $n = 1$. This will mean there are four unknowns A_1, B_1, A_2, B_2 which have to be determined from the boundary and field conditions. We may need more terms (such as $n = 2$) but let us try simplicity first.

Region 1   As $\rho \to \infty$, the effect of the dielectric cylinder must fall off and $V \to -E_0\rho \cos \phi$. Hence $A_1 = -E_0$ and thus

$$V_1 = \left(-E_0\rho + \frac{B_1}{\rho}\right) \cos \phi$$

Region 2 $V_2 = (A_2\rho + B_2/\rho) \cos \phi$. As $\rho \to 0$, the potential V_2 cannot be infinite so that $B_2 = 0$ and we write

$$V_2 = A_2\rho \cos \phi$$

Boundary $\rho = b$ Here E_ϕ (tangential) and D_ρ (normal) must be continuous.

$$E_\phi = -\frac{1}{\rho}\frac{\partial V}{\partial \phi}; \qquad D_\rho = \epsilon E_\rho = -\epsilon \frac{\partial V}{\partial \rho}$$

$$E_{\phi_1} = -\frac{1}{\rho}\left[-\sin\phi\left(-E_0\rho + \frac{B_1}{\rho}\right)\right]$$

$$E_{\phi_2} = -\frac{1}{\rho}(-\sin\phi \, A_2\rho)$$

Equate the last two equations and use $\rho = b$.

$$-E_0b^2 + B_1 = A_2b^2 \qquad\qquad (1)$$

$$D_{\rho_1} = -\epsilon_1 \frac{\partial V_1}{\partial \rho} = -\epsilon_1 \cos\phi \left(-E_0 - \frac{B_1}{\rho^2}\right)$$

$$D_{\rho_2} = -\epsilon_2 \frac{\partial V_2}{\partial \rho} = -\epsilon_2 \cos\phi \, (A_2)$$

Equate the last two equations and use $\rho = b$

$$\epsilon_1(E_0b^2 + B_1) = -\epsilon_2 b^2 A_2 \qquad\qquad (2)$$

From equations 1 and 2 we can determine A_2 and B_1 obtaining

$$B_1 = b^2 E_0 \frac{\epsilon_2 - \epsilon_1}{\epsilon_2 + \epsilon_1} \qquad \text{and} \qquad A_2 = \frac{-2\epsilon_1 E_0}{(\epsilon_2 + \epsilon_1)}$$

Potential solutions are

$$V_1 = E_0 \left(-\rho + b^2 \frac{\epsilon_2 - \epsilon_2}{\epsilon_2 + \epsilon_1} \frac{1}{\rho} \right) \cos \phi$$

$$V_2 = -E_0 \frac{2\epsilon_1}{\epsilon_2 + \epsilon_1} \rho \cos \phi$$

For the field inside the dielectric, $E_2 = -\,\text{grad}\,V_2$, and noting that $x = \rho \cos \phi$ we obtain

$$V_2 = -E_0 \frac{2\epsilon_1}{\epsilon_2 + \epsilon_1} x$$

$$E_2 = -\frac{\partial V}{\partial x} a_x \text{ only} = +E_0 \frac{2\epsilon_1}{\epsilon_2 + \epsilon_1} \text{ V/m}$$

$$D_2 = +\epsilon_0 E_0 \frac{2\epsilon_r}{1 + \epsilon_r} \quad \text{for region 1 air}$$

Thus for $\epsilon_r > 1$, the electric flux density increases in the dielectric. It is uniform there. Figure 9.13b gives a rough indication of the flux and potential field. The lines of electric flux density D crowd into the dielectric and, as there is no free charge on the surface, the D lines are continuous across the boundary and through the dielectric.

9.3.3   Spherical coordinates—Legendre polynomials

Suppose we consider problems which have field symmetry so that, choosing the z axis correctly and using spherical coordinates, the field pattern is independent of ϕ. Then Laplace's equation in spherical coordinates becomes

$$\nabla^2 V = \frac{\partial}{\partial r} \left(r^2 \frac{\partial V}{\partial r} \right) + \frac{1}{\sin \theta} \frac{\partial}{\partial \theta} \left(\sin \theta \frac{\partial V}{\partial \theta} \right) = 0$$

As before, if we assume a solution of the form $V = R(r)T(\theta)$ it should be possible to separate the variables and solve. Substituting

$$T \frac{\partial}{\partial r} \left(r^2 \frac{\partial R}{\partial r} \right) + \frac{R}{\sin \theta} \frac{\partial}{\partial \theta} \left(\sin \theta \frac{\partial T}{\partial \theta} \right) = 0$$

Dividing by RT

$$\frac{1}{R} \frac{\partial}{\partial r} \left(r^2 \frac{\partial R}{\partial r} \right) + \frac{1}{T \sin \theta} \frac{\partial}{\partial \theta} \left(\sin \theta \frac{\partial T}{\partial \theta} \right) = 0$$

The first term is a function of r only and the second is a function of θ. Hence both must be constants, say $\pm K$, so that they sum to zero; hence

$$\frac{1}{R} \frac{\partial}{\partial r} \left(r^2 \frac{\partial R}{\partial r} \right) = K \quad \text{and} \quad \frac{1}{T \sin \theta} \frac{\partial}{\partial \theta} \left(\sin \theta \frac{\partial T}{\partial \theta} \right) = -K$$

Consider one term of a possible power series $R = r^n A_n$

$$K = \frac{1}{r^n A_n} \frac{\partial}{\partial r} \left[r^2 n r^{n-1} \right] A_n = n(n+1)$$

Also if $R = B_n r^{-n-1}$ then

$$K = \frac{1}{B_n r^{-n-1}} \frac{\partial}{\partial r} \left[r^2 (-n-1) r^{-n-2} \right] B_n$$

$$= \frac{1}{r^{-n-1}} (-n-1)(-n) r^{-n-1}$$

$$= n(n+1)$$

Thus if $A_n r^n$ is one solution so is $B_n r^{-(n+1)}$; the solutions arise in pairs and generally for $R(r) = \Sigma R_n$

$$R_n = A_n r^n + B_n r^{-(n+1)} \qquad \text{and} \qquad V = \sum_{0,1}^{\infty} R_n T_n$$

As regards T_n we have

$$\frac{1}{T_n \sin \theta} \frac{\partial}{\partial \theta} \left(\sin \theta \frac{\partial T_n}{\partial \theta} \right) = -K = -n(n+1)$$

$$\frac{1}{\sin \theta} \frac{\partial}{\partial \theta} \left(\sin \theta \frac{\partial T_n}{\partial \theta} \right) + T_n n(n+1) = 0$$

This is really a true differential, now separated.

Write

$$c = \cos \theta \qquad \text{and} \qquad \sin \theta = (1 - c^2)^{1/2}; \qquad dc = (-\sin \theta) \, d\theta$$

Hence

$$\frac{d}{dc} \left(\sin^2 \theta \frac{dT_n}{dc} \right) + n(n+1) T_n = 0$$

on substitution.

$$\frac{d}{dc} \left[(1 - c^2) \frac{dT_n}{dc} \right] + n(n+1) T_n = 0 \quad \text{Legendre's equation}$$

Rearranging

$$(1 - c^2) T_n'' - 2c T_n' + n(n+1) T_n = 0$$

Taking only integer values of n, the finite solutions for T_n are called Legendre polynomials and are in the general form

$$T_n = D_{n0} + D_{n1} c + D_{n2} c^2 + \ldots + D_{nn} c^n$$

From Legendre's equation it can be seen that for integer n the highest order term possible is $D_{nn}c^n$.

For a particular n, the above general polynomial can be inserted in Legendre's equation and the result then arranged in terms of powers of c; each such term can be separately equated to zero, so allowing the $D_{n0}, D_{n1}, \ldots, D_{nn}$ coefficients to be determined.

This is a lengthy process. However, Weber quotes a general expression for T_n normalised

$$T_n = \left(\frac{1.3.5\ldots(2n-1)}{n!}\right)\left\{c^n - \frac{n(n-1)}{2(2n-1)}c^{n-2} \right.$$
$$\left. + \frac{n(n-1)(n-2)(n-3)}{2.4.(2n-1)(2n-3)}c^{n-4} - \ldots + \ldots\right\}$$

Using this, or the particular n process, the first few explicit polynomials can be shown to be as displayed in the table. For the normalised polynomials the D coefficients have been adjusted so that $T_n = 1$ when $c = 1$.

The final solution is

$$V = \sum_{n=0,1,\ldots}^{\infty}(A_n r^n + B_n r^{-(n+1)})T_n$$

	Legendre polynomials	
n	T_n	Normalised T_n
0	$D_0(1)$	1
1	$D_1(c)$	c
2	$D_2(3c^2-1)$	$(1/2)(3c^2-1)$
3	$D_3(5c^3-3c)$	$(1/2)(5c^3-3c)$
4	$D_4(35c^4-30c^2+3)$	$(1/8)(35c^4-30c^2+3)$
5	$D_5(63c^5-70c^3+15c)$	$(1/8)(63c^5-70c^3+15c)$
6	$D_6(231c^6-315c^4+105c^2-5)$	$(1/16)(231c^6-315c^4+105c^2-5)$

Legendre polynomials are used in such problems as

(i) solid or hollow sphere in a uniform field (electric/magnetic)
(ii) field due to ring/disc charged electrically
(iii) field due to a circular current loop
(iv) field in space defined by a conical boundary

Example–hollow magnetic sphere in uniform field H_0 A particularly interesting example of the application of Legendre polynomials is that of a hollow magnetic sphere in a uniform magnetic field as illustrated in figure 9.14. The original components of the magnetic field strength H_0 in the diagram are

$$H_{r0} = H_0\cos\theta a_r; \qquad H_{\theta0} = -H_0\sin\theta a_\theta; \qquad H_{\phi0} = 0$$

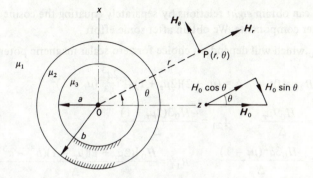

Figure 9.14   *Hollow magnetic sphere in uniform magnetic field.*

and for permeabilities

$$\mu_1 = \mu_3 = \mu_0; \qquad \mu_2 = \mu_r\mu_0$$

By inspection of the field symmetry, we can assume the scalar magnetic potentials U can contain only Legendre terms corresponding to $n = 0$ and $n = 1$. We can write

$$
\left.
\begin{aligned}
b < r \qquad U_1 &= (A_{10} + B_{10}r^{-1})T_0 + (A_{11}r + B_{11}r^{-2})T_1 \\
a < r < b \qquad U_2 &= (A_{20} + B_{20}r^{-1})T_0 + (A_{21}r + B_{21}r^{-2})T_1 \\
r < a \qquad U_3 &= (A_{30} + B_{30}r^{-1})T_0 + (A_{31}r + B_{31}r^{-2})T_1
\end{aligned}
\right\}
\quad
\begin{aligned}
T_0 &= 1 \\
T_1 &= \cos\theta
\end{aligned}
$$

Region 1   As $r \to \infty$, $U \to 0$ so that A_{10} and A_{11} both equal zero.

Region 3   As $r \to 0$, U cannot go infinite so B_{30}, B_{31} are both zero. Thus

$$U_1 = B_{10}r^{-1} + B_{11}r^{-2}\cos\theta$$

$$U_2 = (A_{20} + B_{20}r^{-1}) + (A_{21}r + B_{21}r^{-2})\cos\theta$$

$$U = A_{30} + A_{31}r\cos\theta$$

These are the three equations for scalar potential.

Boundary $r = b$ The magnetic field strength H_θ and magnetic flux density B_r must be continuous

$$H_{\theta 0} - \frac{1}{r}\frac{\partial U_1}{\partial\theta} = -\frac{1}{r}\frac{\partial U_2}{\partial\theta}; \qquad \mu_1\left(H_{r0} - \frac{\partial U_1}{\partial r}\right) = -\mu_2\frac{\partial U_2}{\partial r}$$

Boundary $r = a$ Again H_θ and B_r must be continuous

$$-\frac{1}{r}\frac{\partial U_2}{\partial\theta} = \frac{-1}{r}\frac{\partial U_3}{\partial\theta}; \qquad -\mu_2\frac{\partial U_2}{\partial r} = -\mu_3\frac{\partial U_3}{\partial r}$$

There are eight coefficients to be determined in the three potential equations and we have only *four* boundary equations. But since these latter must be true for all

θ values we can obtain *eight* relations by separately equating the cosine terms and the other components. We obtain after some effort

A_{20}, A_{30} which will depend on choice for zero scalar magnetic potential U

$$B_{10}, B_{20} = 0; \qquad \Delta = (\mu_r + 2)(2\mu_r + 1) - \frac{2a^3}{b^3}(\mu_r - 1)^2$$

$$A_{31} = \frac{-H_0 9\mu_r}{\Delta}; \qquad A_{21} = \frac{-H_0 3(2\mu_r + 1)}{\Delta}$$

$$B_{21} = \frac{-H_0 3a^2(\mu_r - 1)}{\Delta}; \qquad B_{11} = \frac{H_0\,[(2\mu_r + 1)(\mu_r - 1)(b^3 - a^3)]}{\Delta}$$

Now suppose we want to find the field in the iron and in the hollow centre.

$$H_3 = -\operatorname{grad} U_3 = -\nabla(A_{30} + A_{31}r\cos\theta) = -\nabla(A_{30} + A_{31}z)$$

$$= -A_{31}a_z = \frac{H_0 9\mu_r}{(\mu_r + 2)(2\mu_r + 1) - (2a^3/b^3)(\mu_r - 1)^2}\, a_z$$

This is a *uniform field* in direction a_z, as was H_0. If μ_r is large, this will be less than the original H_0, so that the shielding factor is

$$S = \frac{H_0}{H_3} = \frac{(\mu_r + 2)(2\mu_r + 1) - (2a^3/b^3)(\mu_r - 1)^2}{9\mu_r}$$

If the shell is thick so that $(a/b)^3$ is small, the second term in the numerator can be ignored, giving

$$S = \frac{H_0}{H_3} \approx \frac{2\mu_r^2 + 5\mu_r + 2}{9\mu_r} \approx \frac{2\mu_r + 5}{9}$$

Thus for a material of $\mu_r = 11$, S is as high as 3 and the field in the inside of the shell has fallen to $\frac{1}{3}H_0$.

We can determine the field in the iron from U_2. It will have both H_r and H_θ components

$$H_r = -\frac{\partial U_2}{\partial r} = -\left(A_{21} - \frac{2B_{21}}{r^3}\right)\cos\theta = \frac{-\cos\theta}{\Delta}\left\{-3H_0(2\mu_r + 1) + \frac{2}{r^3}H_0 3a^3(\mu_r + 1)\right\}$$

$$H_\theta = -\frac{1}{r}\frac{\partial U_2}{\partial\theta} = \frac{1}{r}\sin\theta\left(A_{21}r + \frac{B_{21}}{r^2}\right) = \frac{\sin\theta}{\Delta}\left\{-3H_0(2\mu_r + 1) - \frac{H_0}{r^3}3a^3(\mu_r + 1)\right\}$$

From these we can find the *field in a solid sphere* by making $a \to 0$, giving

$$H = H_r a_r + H_\theta a_\theta = \frac{3H_0(2\mu_r + 1)}{\Delta}(\cos\theta a_r - \sin\theta a_\theta)$$

$$= \frac{3H_0(2\mu_r + 1)}{\Delta}a_z = \frac{3H_0}{\mu_r + 2}a_z$$

that is a uniform field. The field in a solid cylinder is unchanged if $\mu_r = 1$, and then falls off practically inversely as μ_r. This of course applies to the magnetic field strength H. For example, for material of $\mu_r = 10$, H falls to $H_0/4$ but the magnetic flux density B rises from B_0 to $(5/2)B_0$, in the iron.

For the *field external to the sphere* this can be found from $-\text{grad } U_1$, remembering to add H_0 the applied field strength; there will be H_r and H_θ components. It can be shown that near the shell at $\theta = 0$, $H \to H_0 3\mu_r/(\mu_r + 2)$. Near a ship in earth's field, the field strength can increase by a factor of 3.

Problems

9.1 The yz plane is a conductor. A charge Q is assumed to be uniformly distributed on a conducting sphere of radius a whose centre is placed at $(b, 0, 0)$ and $b \gg a$. Show that the force on the sphere is $(Q^2/4\pi\epsilon_0)(1/4b^2)$ and the maximum charge density on the yz conducting plane is $-Q/2\pi b^2$. How would the quantities be affected if a parallel conducting plane were erected through $x = 2b$?

$(0, (-0\cdot92)Q/2\pi b^2)$

9.2 Find the mechanical force (N/m) on a long, straight wire carrying 1000 A d.c. at space coordinates $(0\cdot03, 0\cdot04, z)$ for the following *four* cases.

(a) Region $(x \geqslant 0, y \geqslant 0, z)$ is air, the rest is iron of high permeability.
(b) Region $(x \geqslant 0, y \geqslant 0, z)$ is iron of high permeability, the rest is air.
(c) As (a) but using $\mu_r = 9$.
(d) As (b) but using $\mu_r = 9$.
((a) $6\cdot11$ N/m; (b) $2\cdot32$ N/m; (c) $5\cdot48$ N/m; (d) $2\cdot13$ N/m)

9.3 (a) Draw the curves $u = $ constant and $v = $ constant for conformal transformation $u + jv = (x + jy)^{1/2}$. Use values of u and v of 0, 0·5, 1·0, 1·5 and take x and y over range -5 to $+5$. Use scale 1 unit $= \frac{1}{2}$ inch or 1 unit $= 1$ cm. Suggest the field problem represented.
(b) In a two-dimensional field problem the flow function $u = x^2 - y^2$. Find the potential function v. Draw the u, v curves for unit steps, confining attention to the first quadrant in range 0 to 3 for x, y. What practical problem is represented?

9.4 A long, rectangular plate, thickness t, conductivity σ, width π, lies in the xy plane and has its four edges defined by $x = 0$, $x = \pi$, $y = 0$ and $y = $ large and positive. Three edges are grounded but the $y = 0$ edge is held at a fixed potential V_0. Two corners are removed by cutting along flow lines passing through the $y = 0$ edge at $x = \pi/10$ and at $x = 9\pi/10$. Find the conductance from the $y = 0$ edge to ground.

$(2\cdot34\sigma t)$

9.5   A hollow cylinder of inner, outer radii a, b, of iron of high μ_r, is placed in and orthogonal to an extensive uniform field of strength H_0. Calculate the magnetic field strength H_3 inside the cylinder and show that the shielding factor $S = H_0/H_3 = (\mu_r/4)[(1 - (a/b)^2]$.

      If $\mu_r = 200$ and a/b is of order $4/5$, what would you suggest to obtain S about 300 at the cylinder centre?

9.6   A conducting sphere of radius a carries a charge Q and is surrounded by a shell of material of dielectric constant ϵ_r. The outer radius of the shell is b. Calculate the potential at the following points: (i) outside the dielectric, (ii) inside the dielectric, (iii) at the surface of the sphere.

$$\left(\text{(i)} \ \frac{Q}{4\pi\epsilon_0 r} \qquad \text{(ii)} \ \frac{Q}{4\pi\epsilon_0 r} \left(\frac{1}{r} - \frac{\epsilon_r - 1}{b} \right) \qquad \text{(iii)} \ \frac{Q}{4\pi\epsilon_0 \epsilon_r} \left(\frac{1}{a} - \frac{\epsilon_r - 1}{b} \right) \right)$$

9.7   A spherical cavity of radius a is cut from an infinite dielectric medium with dielectric constant ϵ_r. An electric dipole of moment p directed along the z axis is placed at the centre of the cavity. Calculate the potentials inside and outside the cavity.

$$\left(\frac{p \cos \theta}{4\pi\epsilon_0 r^2} \left[1 - \frac{2(\epsilon_r - 1)}{2\epsilon_r + 1} \frac{r^3}{a^3} \right], \frac{3p \cos \theta}{4\pi\epsilon_0 (2\epsilon_r + 1) r^2} \right)$$

9.8   A voltage V is applied to a thin parallel-plate capacitor of plate separation d which is filled with a uniform space charge of density ρ_v. If the potential of the positive plate is V_0 calculate (i) the potential between the plates, (ii) the electric field between the plates, (iii) the surface charge densities on the plates.

$$\left(\text{(i)} \ -\frac{\rho_v x^2}{2\epsilon_0} + \left(\frac{\rho_v d}{2\epsilon_0} - \frac{V}{d} \right) x + V_0; \qquad \text{(ii)} \left(\frac{V}{d} + \frac{\rho}{2\epsilon_0} (2x - d) \right) a_x; \right.$$

$$\left. \text{(iii)} \ \frac{\epsilon_0 V}{d} - \frac{\rho_v d}{2}, \frac{-\epsilon_0 V}{d} - \frac{\rho_v d}{2} \right)$$

9.9   By direct integration of Laplace's equation find the potential at all points within a coaxial cylindrical cable which has an inner radius a at V_0 volts and an outer of radius b. Find the electric field strength.

$$\left(\frac{V_0 \ln (b/r)}{\ln (b/a)}, \frac{V_0}{r \ln (b/a)} a_\rho \right)$$

9.10 A pair of coaxial conducting cylinders of radius a and c are separated by two coaxial dielectrics. One dielectric between radii a and b has dielectric constant ϵ_{r_1} the other between radii b and c has dielectric constant ϵ_{r_2}. The potential of the outer cylinder is V_0 with respect to the inner cylinder which is earthed. Using the appropriate solutions to Laplace's equation derive the potentials within the dielectrics.

$$\left(\frac{V_0 \ln (r/a)}{\ln (b/a) + (\epsilon_{r_1}/\epsilon_{r_2}) \ln (c/a)}; \quad V_0 \left\{ 1 - \frac{(\epsilon_{r_1}/\epsilon_{r_2}) \ln (c/r)}{\ln (b/a) + (\epsilon_{r_1}/\epsilon_{r_2}) \ln (c/b)} \right\} \right)$$

10

Some Low-frequency Applications

In the first nine chapters of this text we have developed concepts of electromagnetism and Maxwell's equations and have attempted to discuss applications of these in situations of direct practical concern. It should now be obvious that the theories of electromagnetism have use throughout the whole field of electrical and electronic engineering. In this chapter we shall consider a few low-frequency applications and the problems we have chosen to discuss are governed by the authors' separate, rather than mutual, interests.

The first example we shall choose will be from the field of electrical machines and we begin by considering Maxwell's equation for curl E

$$\text{curl } E = -\frac{\partial}{\partial t} B = -\dot{B}$$

In integral form this may be written

$$\oint_l E \cdot dl = -\int_S \frac{\partial B}{\partial t} \cdot dS$$

and is interpreted as Faraday's induced voltage relationship

$$\text{voltage induced in loop, } V_{\text{emf}} = -\frac{\partial \phi}{\partial t}$$

$$= -\text{rate of change of flux embraced by the loop}$$

In this form the equation is easy to understand and may be compared with the equation of curl H which is written in integral form

$$\oint_l H \cdot dl = \int_S J \cdot dS + \int_S \frac{\partial D}{\partial t} \cdot dS$$

196

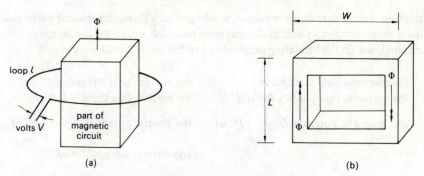

Figure 10.1 *Part and complete magnetic circuit.*

This equation has been discussed in some detail in section 6.2.1 in connection with the concept of displacement current. The electric and magnetic effects of currents may be obtained from these equations but if ferromagnetic material is present then the equation div $B = 0$ must also be applied.

A portion of a magnetic circuit is shown in figure 10.1a and the complete magnetic circuit in figure 10.1b. The equation for curl E states that, in the space outside the ferromagnetic core which contains the changing flux $d\Phi/\partial t$, there is an electric stress pattern.

Should the complete circuit be such that W is finite and L infinite then the field pattern is given in figure 10.2.

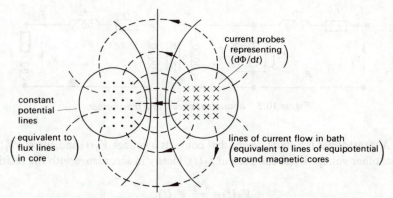

Figure 10.2 *Field pattern for transformer core.*

The pattern has already been considered in section 9.2.2. Should a closed copper winding be placed around the core then the changing flux induces a current in it; this modifies both the flux pattern and the electric stress pattern.

10.1 General observations about Maxwell's equations

From the foregoing it can be seen that in any electrical machine, such as a transformer, motor or generator, *all* Maxwell's equations are involved. Their complete solution, everywhere in the machine, would give us a three-dimensional pattern of the utmost complexity within the magnetic material, within the conducting material, within the insulating material and in the air. It would involve

the magnetic field strength H	the electric field strength E
the magnetic flux density $B = \mu H$	the electric flux density $D = \epsilon E$
and magnetic potential $U = - \int_l H \cdot \mathrm{d}l$	the electric potential $V = - \int_l E \cdot \mathrm{d}l$
	and current density $J = \sigma E$

10.2 Maxwell's equations applied to a transformer

As far as is known, *no one* has attempted to solve Maxwell's equations in complete detail for the relatively simple static transformer. Fortunately we are not normally interested in knowing about everything that is going on and for the transformer an 'equivalent circuit' is a substitute to account for the macroscopic performance (relations between voltages V_1, V_2, currents I_1, I_2 and load impedance Z) and it avoids the interminable detail of field theory (Maxwell's equations, polarisation, domain theory of ferromagnetism, etc.). The simple equivalent circuit is shown in figure 10.3.

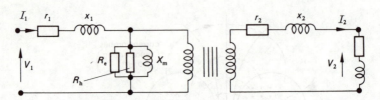

Figure 10.3   *Equivalent circuit of a transformer.*

A transformer is a static device that converts a voltage $V_1(t)$ and current $I_1(t)$ to another voltage $V_2(t)$ and current $I_2(t)$, ideally in accordance with the relations

$$V_2(t) = \frac{N_2}{N_1} V_1(t)$$

$$I_2(t) = \frac{N_1}{N_2} I_1(t)$$

where N are the turns in the appropriate windings; the subscripts 1 and 2 refer to the primary and secondary respectively.

Notice that V_1 is a certain function of time and ideally V_2 should be of similar form, differing only in magnitude. Likewise I_2 and I_1 should have similar time functions, differing again in magnitude.

There are many types of transformer.

(1) Voltage transformers with sinusoidal time functions;
(2) Audio frequency voltage transformers, having a spectrum of frequencies present, together or at differnt times;
(3) Pulse voltage transformers where the voltage time function is a pulse ⊓;
(4), (5), (6) Current transformers for sinusoidal, audio-frequency and pulse operation;
(7) Rectifier transformers (carry mixed a.c. and d.c.).

The transformers can be of 50 Hz for transforming thousands of MVA of power between voltages from 440 V to 400 kV. They can also be for ratings down to miniature hearing-aid transformers, and transformers for micro-minature circuits. They can cover a wide frequency range from 1 Hz, to power frequency (50, 60 Hz) to audio frequencies (up to 15 kHz) and to radio frequencies (well into the megahertz range).

It is no exaggeration to state that Maxwell himself would have difficulty in coping with the electromagnetic detail of the transformer which is in fact almost the simplest electromagnetic machine. Take the one aspect of eddy currents in particular.

10.2.1 The eddy-current problem in the transformer

Transformers are almost always made of laminated steel cores; this excludes those very high frequency ferrite cores, and cores for very low frequency and very low power. Maxwell's equations point out the reasons for this.

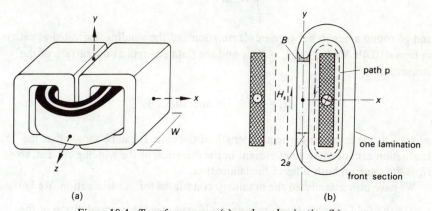

Figure 10.4 *Transformer core* (a), *and one lamination* (b).

Imagine a core built up from laminations in the conventional form of C cores, as we have indicated in figure 10.4a. To simplify the problem assume the dimension W is infinite to reduce the problem to one of two dimensions. Consider a single lamination, a magnetising winding and an unloaded secondary, shown in figure 10.4b with the lamination encircling the right-hand winding.

If a voltage V_1 is applied to the winding of N_1 turns then if we take a circuit (loop) l_w formed by the winding itself

$$\oint_{l_w} E . \, dl = -\frac{\partial}{\partial t} \int_S B . \, dS$$

which is interpreted more familiarly as

$$\frac{\text{volts}}{\text{turn}} = \frac{V_1}{N_1} = A \frac{dB_{mean}}{dt}$$

where B_{mean} is the mean magnetic flux density. If we assume that all the laminations carry the same flux density, the mean flux density in our lamination is forced to the value given by

$$\frac{dB_{mean}}{dt} = \frac{V_1}{N_1 A}$$

in which A is the total core area. If now $V_1 = \hat{v}_1 \sin(\omega t)$ we can write

$$B_{mean} = \int \frac{\hat{v}_1 \sin(\omega t)}{NA} \, dt = -\frac{\hat{v}_1}{N_1 A \omega} \cos(\omega t)$$

Now, assuming that the magnetising winding carries a current I, we can apply Maxwell's equation

$$\oint_l H . \, dl = I$$

and go round a circuit between our lamination and the winding, indicated as path p in figure 10.4b. We have $H_s l_c = N_1 I_1$ and the field strength at the surface of the lamination is

$$H_s = \frac{N_1 l_1}{l_c} \quad (\text{in direction } Y)$$

where l_c is the mean magnetic path length of the core. It can be shown that the lamination cannot carry a net current in the direction of the winding current, so H_s is the same on both sides of the lamination.

We have now established the boundary conditions for our lamination. We know

(i) the field strength H_s at its surfaces. If we knew the permeability μ of the material we would also know the flux density at the surface $B_s = \mu H_s$.

(ii) The average value of flux density across the lamination thickness

$$B_{mean} = \frac{1}{a} \int_0^a B \, dx = \frac{\hat{v}_1}{N_1 A \omega} \cos(\omega t)$$

What we do not know is how the field strength H and flux density B are distributed across the lamination thickness. Now look at the lamination in plane,

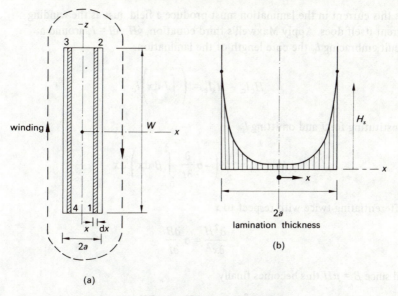

Figure 10.5 *Field conditions in a lamination.*

the y direction in figure 10.5a. Suppose width $W \gg$ thickness $2a$. Consider an elementary thickness dx, distance x from the lamination centre-line and on either side; that is the circuit shown as 1, 2, 3, 4.

Maxwell's second equation says

$$E \cdot dl = -\frac{\partial}{\partial t} \int_S B \cdot dS$$

The voltage between 2, 3 (and 4, 1) is much less than that between 1, 2 (and 3, 4). Assume that B, which is in the y direction, is a function of x only, and E is a constant along the element; this is reasonable from the symmetry. Then

$$2WE = -\frac{\partial}{\partial t} \int_{-x}^{+x} (B \, dx)W$$

or

$$E = -\frac{\partial}{\partial t} \int_0^x B \, dx \text{ (in the } z \text{ direction)}$$

This electric field will give rise to a current, of density J, in the lamination

$$E = J/\sigma$$

where σ is the conductivity of the lamination material. Hence

$$\frac{J}{\sigma} = -\frac{\partial}{\partial t} \int_0^x B \, dx$$

But this current in the lamination must produce a field, just as the winding current itself does. Apply Maxwell's third equation, $\oint H \cdot dl = I$, around a circuit embracing l_c the core length of the lamination.

$$H_c l_c - H_s l_c = \left(\int\limits_x^a J \, dx \right) l_c$$

Substituting for J and omitting l_c

$$H - H_s = \int\limits_x^a \left(-\sigma \frac{\partial}{\partial t} \int\limits_0^x B \, dx \right) dx$$

Differentiating twice with respect to x

$$\frac{\partial^2 H}{\partial x^2} = \sigma \frac{\partial B}{\partial t}$$

and since $B = \mu H$ this becomes finally

$$\frac{\partial^2 H}{\partial x^2} = \mu\sigma \frac{\partial H}{\partial t} = \kappa^2 \frac{\partial H}{\partial t} \qquad\qquad \text{diffusion equation}$$

The diffusion equation is a known equation; it is sometimes called the 'eddy-current' or 'skin-effect' problem. It was first derived by Fourier in solving heat flow problems; in these, κ^2 is the thermal conductivity σ_{th} divided by the volume specific heat c_v and κ is known as the 'diffusivity'.

To solve this differential equation we must apply the known boundary conditions. These are

$$\frac{\partial H}{\partial x} = 0 \qquad \text{at} \qquad x = 0 \ \text{(from symmetry)}$$

and *either*

$$H_a = H_s = \frac{N_1 \hat{i}_1}{l_c} \sin(\omega t)$$

if the magnetising current I_1 is known, *or*

$$\frac{1}{a} \int\limits_0^a B \, dx = -\frac{\hat{v}_1}{N_1 A \omega} \cos(\omega t)$$

if the voltage V_1 across the magnetising coil is known. Finally

$$\frac{H}{H_s} = \frac{C'c' + jS's'}{Cc + jSs}$$

where

$$C' = \cosh \alpha x, \qquad c' = \cos \alpha x, \qquad S' = \sinh \alpha x, \qquad s' = \sin \alpha x$$
$$C = \cosh \alpha a, \qquad c = \cos \alpha a, \qquad S = \sinh \alpha a, \qquad s = \sin \alpha a$$

and

$$\alpha = \kappa (\omega/2)^{1/2} = (\pi \mu \sigma f)^{1/2}$$

A graph of this relation is shown in figure 10.5b.

As the frequency increases the eddy currents (described by the current density J) prevent the flux from penetrating to the centre of the lamination. The eddy-current losses increase and the effective inductance drops since a significant proportion of the core is virtually unused.

The depth of penetration of flux is defined as

$$d_p = (\pi \mu f \sigma)^{-1/2}$$

If the lamination thickness is significantly greater than $2d_p$ the core material is not being properly used. This is obviously of the same form as the skin effect in section 7.5.4.

10.3   Diffusion equation

It *is* possible to base the derivation on the differential forms of Maxwell's equations in a purely mathematical approach without appreciating the physical processes involved.

Assume μ, ϵ are non-directional and constant. Then

$$\nabla \times E = -\mu \frac{\partial H}{\partial t}$$

$$\nabla \times H = \sigma E + \epsilon \frac{\partial E}{\partial t}$$

Now try to eliminate one of the two unknown variables E, H. Take the curl of the first, then substitute the second

$$\nabla \times (\nabla \times E) = -\mu \frac{\partial}{\partial t} \nabla \times H$$

$$= -\mu \frac{\partial}{\partial t} \left(\sigma E + \epsilon \frac{\partial E}{\partial t} \right)$$

Make use of the identity curl curl = grad div − lap

$$\nabla (\nabla . E) - \nabla^2 E = -\mu \sigma \frac{\partial E}{\partial t} - \mu \epsilon \frac{\partial^2 E}{\partial t^2}$$

Now introduce Maxwell's first equation which says

$$\nabla . D = \nabla . \epsilon E = \epsilon \nabla . E = \rho_V$$

In free space, $\rho_V = 0$ and in a conductor (the lamination) it is independent of the distribution of electric field strength E, so $\nabla . E = 0$ eliminating the 'grad div' term

$$-\nabla^2 E = -\mu\sigma \frac{\partial E}{\partial t} - \mu\epsilon \frac{\partial^2 E}{\partial t^2}$$

and inserting $J = \sigma E$ this becomes

$$\nabla^2 J = \mu\sigma \frac{\partial J}{\partial t} + \mu\epsilon \frac{\partial^2 J}{\partial t^2}$$

If, instead of eliminating H from the two 'curl equations' at the start, we had eliminated E, the result would have been

$$\nabla^2 H = \mu\sigma \frac{\partial H}{\partial t} + \mu\epsilon \frac{\partial^2 H}{\partial t^2}$$

There are *two* important instances.

(i) In free space $\sigma = 0$ and the wave equation is produced

$$\nabla^2 H = \mu\epsilon \frac{\partial^2 H}{\partial t^2} \qquad \text{wave equation}$$

(ii) In metals at frequencies up to 10^3, 10^4 MHz, the second term is negligible, $\epsilon_0 \approx 1/36\pi 10^9$ is dominant, and the diffusion equation is obtained

$$\nabla^2 H = \mu\sigma \frac{\partial H}{\partial t} \qquad \text{diffusion equation}$$

10.4 Field produced by an element of changing flux

The current element $i\,dl$ is used as a basis for computing the magnetic field strength at a point distant r by

$$dH = i\,dl \times \frac{a_r}{4\pi r^2}$$

Is there a 'magnetic element' which can be used to compute the electric field strength in the same way? We suspect that there must be, so let us proceed thus.

Replace the current element $i\,dl$ by a volume element $J\,d(\text{vol})$, see figure 10.6, then

$$dH = J\,d(\text{vol}) \times \frac{a_r}{4\pi r^2}$$

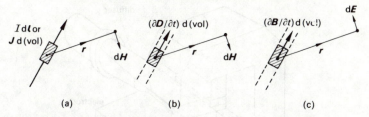

Figure 10.6   *Field produced by volume element of changing flux.*

Now consider Maxwell's third equation for curl H

$$\nabla \times H = J + \frac{\partial}{\partial t} D$$

It is clear from this that the term $(\partial/\partial t)D$ for the changing electric flux density may be related to a magnetic effect in just the same way as J. So for a volume element in which $(\partial/\partial t)D$ occurs there will be produced a magnetic field

$$\mathrm{d}H = \frac{\partial}{\partial t} D \, \mathrm{d}(\mathrm{vol}) \times \frac{a_r}{4\pi r^2}$$

Turn now to Maxwell's second equation for curl E

$$\nabla \times E = -\frac{\partial}{\partial t} B$$

By inference the electric field strength $\mathrm{d}E$ produced by a volume element in which there is a changing magnetic field will be as shown in figure 10.6c.

$$\mathrm{d}E = -\frac{\partial}{\partial t} B \, \mathrm{d}(\mathrm{vol}) \times \frac{a_r}{4\pi r^2}$$

This is the 'volume-element form' of Maxwell's second equation and is useful for certain devices such as particle accelerators.

10.5   The electromagnetic pump

This is an example drawn from the growing field of magnetohydrodynamics (or magneto-fluid-mechanics) which concerns conducting fluids moving in magnetic fields.

In a normal electric motor, the torque on the rotor results from the interaction of the current in the rotor bars and the magnetic field produced by the stator coils. The directions of field, current and mechanical force are related by Fleming's left-hand rule.

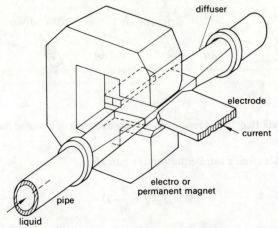

Figure 10.7 *Simple conduction pump.*

The electromagnetic pump operates on the same principle. If the liquid to be pumped is a conductor, it can be made to carry a current I. If a magnetic field is established perpendicular to I, the liquid itself will experience a force F perpendicular to both the magnetic field strength H and I as in the motor. Thus conducting liquid can be pumped without the need for a mechanical impeller.

The d.c. conduction pump in its simplest form as illustrated in figure 10.7 is based on this principle. Notice that the liquid flows in a pipe so that glands and seals, required by an impeller, are avoided. This can be a real advantage with liquids such as sodium or sodium–potassium which are chemically reactive in the presence of air and moisture and could also be radioactive if used as a nuclear-reactor coolant.

As with electric motors, there are several types of electromagnetic pump, depending on the following

(1) how the current is established in the liquid metal, that is from an external power source (conduction types) or induced by the field (induction types),
(2) the kind of current (a.c. or d.c. types),
(3) the geometry of the pump, for example linear or spiral flow of liquid metal.

As with the conductors in a normal generator, a conducting liquid flowing across a magnetic field will generate a voltage and produce current and power. The main use of this principle is the electromagnetic flowmeter. For example, the flow in the Straits of Dover has been estimated from the measured e.m.f. produced by the movement of the sea in the earth's magnetic field.

10.5.1 The conduction pump

Referring to the section of the simple electromagnetic pump shown in figure 10.8, the pressure produced in the liquid is, if the magnetic flux density is B

$$p = JBc$$

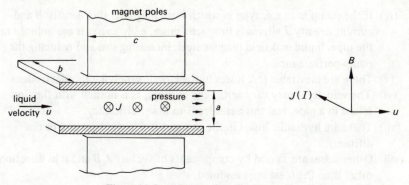

Figure 10.8 *Section of conduction pump.*

in which it is assumed that the current density $J\,(\mathrm{A/m^2})$ is everywhere constant over field length c.

The voltage V_E required across the electrodes to establish the current density J is

$$V_E = \left(\frac{1}{\sigma} J + uB\right)b$$

in which σ is the liquid conductivity, u its velocity and b the field breadth.

The pump output is the product of pressure p and flow q

$$p_0 = pq = puab = JBc\,uab$$

in which a, b are the channel dimensions and c the 'active' length.

The input power to the liquid is

$$P_i = V_E I = \left(\frac{1}{\sigma} J + uB\right)bJac$$

and the pump 'theoretical' efficiency is

$$\frac{P_o}{P_i} = \frac{JBuabc}{[(1/\sigma)J + uB]Jabc} = \frac{\sigma uB}{J + \sigma uB}$$

From this, the larger σuB is compared with J, the greater is the efficiency.

This has oversimplified the situation to illustrate its basic principles. In practice there are other losses and factors to consider which considerably lower the pump efficiency. These can be listed as follows.

(i) I^2R losses in the containing pipe walls if, as is usual, the pipe is an electrical conductor.

(ii) The voltage V_E can also produce a current outside the 'active region c' and to minimise this the field B has to be graded to match.

(iii) The current density J produces its magnetic field which distorts the main imposed field of magnetic flux density B; this is a form of 'armature reaction' and leads to increase in loss.

(iv)   If the pump is an a.c. type in which both magnetic flux density B and current density J alternate in synchronism, eddy currents are induced in the pipe, liquid and field magnet steel, increasing loss and reducing the pump performance.

(v)   There are inevitable I^2R losses in the circuit supplying the electrodes.

(vi)   The velocity u may vary across the channel, as is normal with the flow of liquid in a pipe, and this again leads to lower efficiency.

(vii) There are hydraulic losses in the channel and in the inlet and outlet diffusers.

(viii) Other losses are caused by components of vectors J, B and u in directions other than the ideal ones assumed.

10.5.2   Example on the design of a conduction pump

Design a d.c. conduction pump for bismuth at 500 °C to produce a pressure of 60 lb/in² and a flow of 10 gal/min.

Assume a rectangular channel, dimensions a, b of 0·673 cm, 3·68 cm, and effective length c of 7·93 cm; the walls are 0·0635 cm thick and of steel of resistivity 93×10^{-8} Ωm.

Assume hydraulic friction losses of 40 watts. The series winding of 3 turns required to establish the magnetic field across the channel has a resistance of 60$\mu\Omega$. The effective gap d between the poles is 1·0 cm.

The equivalent circuit of the d.c. conduction pump can be represented as shown in figure 10.9.

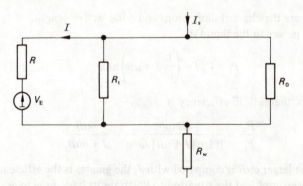

Figure 10.9   *Equivalent circuit of d.c. conduction pump.*

The liquid metal between the electrodes can be assumed to have two resistance components. R is that due to the portion within the magnetic field and R_0 is that due to metal on either side of 'c' and not within the effective field. Also R_t is the effective resistance of the tube walls and R_w the resistance of the field winding. The calculation of these terms based on resistivity times path length divided by section area carrying current needs modification for fringing effects and for 'wetting effects' of the liquid and the channel wall.

Typical values carefully computed are $R = 90\,\mu\Omega$ and $R_0 = 600\,\mu\Omega$, $R_t = 225\,\mu\Omega$. Assume great care has been taken to taper off the magnetic field on either side of 'c' in such a way that the current density outside the main pole area is effectively utilised.

Output power of pump (product of pressure and flow)

$$P_o = \frac{(\text{gal/min})(\text{lb/in}^2)}{1430} \text{ h.p.}$$

$$= \frac{(\quad\cdot\quad)(\quad\cdot\quad)}{1940} \text{ kW}$$

$$= \frac{(10\,\text{gal/min})(60\,\text{lb/in}^2)}{1940} \text{ kW}$$

$$= 0\cdot310 \text{ kW}$$

To this add the 40 W hydraulic loss, giving the electrical output

$$P_o' = 0\cdot350 \text{ kW}$$

$$H = \frac{NI}{d} = \left(\frac{3I_s}{0\cdot01}\right) = 300I_s \text{ A/m}$$

The value of the effective gap distance d has been adjusted to allow for the m.m.f. drop in the iron of the magnet and pipe.

The liquid velocity u depends on flow q and section area ab

$$u = \frac{\text{gal/min}^{-1}}{\text{cm}^2}\,0\cdot755 \text{ m/s}$$

$$= \frac{10}{0\cdot673 \times 3\cdot68}\,0\cdot755 = 3\cdot05 \text{ m/s}$$

$$V_E = Bbu = \mu_0 Hbu = \frac{4\pi}{10^7}\,300I_s\,\frac{3\cdot68}{100}\,3\cdot05$$

$$= 42\cdot3 \times 10^{-6}I_s \text{ V}$$

Employing Thévenin's theorem on the equivalent circuit it is a relatively easy matter to show that

$$\frac{I}{I_s} = \left(R_p - \frac{V_E}{I_s}\right)\bigg/\left(R_p + R\right)$$

In this case R_p is the resistance of R_0, R_t in parallel, or $R_0 R_t/(R_0 + R_t)$ which is $164\,\mu\Omega$; so that working in micro-ohms

$$\frac{I}{I_s} = \frac{164 - 42\cdot3}{164 + 90} = 0\cdot479$$

The electrical output $V_E I = P'_o = 350$ W

$$42 \cdot 3 \times 10^{-6} I_s (0 \cdot 479 I_s) = 350$$

From these

$$I_s = 4156 \text{ A}, \qquad I = 1990 \text{ A}; \qquad V_E = 0 \cdot 1758 \text{ V} \qquad \text{and} \qquad V = 0 \cdot 355 \text{ V}$$

Total input power $P_t = (V_E + IR)I_s + I_s{}^2 R_w$

$$= (0 \cdot 176 + 0 \cdot 179) \, 4156 + 4 \cdot 156^2 \times 60$$

$$= 1475 + 1035 = 25 \cdot 0 \text{ W}$$

Efficiency $\eta = \dfrac{P_o}{P_t} = \dfrac{310}{2510} = 12 \cdot 35$ per cent

10.5.3   Some observations on the conduction pump

Owing to the relatively poor electrical conductivity of bismuth, $0 \cdot 715 \times 10^6$ S/m at 500 °C, 'armature reaction' effects are negligible. If the liquid were sodium, which has a conductivity of $6 \cdot 65 \times 10^6$ S/m at 250 °C, armature reaction would be significant and require the use of compensating bars in the pole faces to prevent large variations in magnetic flux density B and current density J over active length c. Although the ratio $\sigma u B$ to J would be higher, theoretically leading to higher efficiency, the losses in the tube and sides, represented by R_t and R_0, would be correspondingly higher.

In general, a.c. and d.c. conduction pumps are suitable for low-conductivity liquid metals of high density such as bismuth and mercury. With sodium the greater conductivity and higher flow velocity possible permit the use of induction pumps which are simpler to manufacture and do not require the inconvenient high-current low-voltage source (4156 A, 0·355 V in the example above).

The above example is given only as a reference. Maxwell's equations are not much in evidence in this simplified treatment and a full design study for a bismuth conduction pump would attempt to answer the following.

(i)   What is the optimum liquid velocity for high efficiency?
(ii) What are the effects of changing the channel working-dimensions a, b, c?
(iii) If 25 per cent efficiency were essential, can it be achieved and if so what would be the size and shape of the pump?
(iv) What are the final pressure–flow characteristics of the pump?

The problem 10.5 is a typical 'student's exercise' which gives an efficiency far higher than for any practical pump.

10.6   Drift and diffusion in semiconductors

In section 3.3 we introduced the concept of an electron drift velocity in an applied electric field. Thus if an electric field E_x exists in the x direction, the current density J_x for a semiconductor, in which both electrons and holes are found, is

$$J_x = \sigma E_x$$

and

$$\sigma = |q_e|(n\mu_n + p\mu_p)$$

where n and p are the electron and hole charge carrier densities and μ_n and μ_p the respective mobilities. This is not the only current flow possible in a semiconductor, for unlike metals these materials can sustain charge-carrier concentration gradients which give a further component to current flow. Consider figure 10.10 which shows concentration gradients in one dimension and indicates the particle flow and resultant current flow due to a charge-carrier concentration gradient.

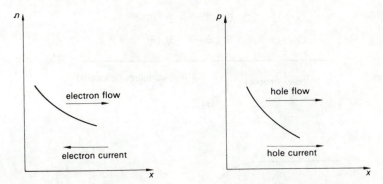

Figure 10.10   *The difference between particle flow and current.* (a) *Electron diffusion.* (b) *Hole diffusion.*

In each case the particle flow is proportional to the concentration gradient at the particular value of x, and the electron and hole currents are $+q_e D_n (\mathrm{d}n/\mathrm{d}x)$ and $-q_e D_p (\mathrm{d}p/\mathrm{d}x)$ where D_n and D_p are the diffusion constants for electrons and holes. The total current flow due to drift and diffusion may be expressed as

$$J_{xn} = q_e \mu_n n E_x + q_e D_n \frac{\mathrm{d}n}{\mathrm{d}x} \qquad \text{for electrons}$$

and

$$J_{xp} = q_e \mu_p p E_x - q_e D_p \frac{\mathrm{d}p}{\mathrm{d}x} \qquad \text{for holes}$$

Note that diffusion takes place in many other environments, not necessarily in materials; examples include atmospheric turbulence, Brownian movement, electrolytic diffusion and atomic diffusion. Whenever a concentration gradient exists then diffusion may well be possible. Diffusion does *not* need an electric field.

10.7   The abrupt *p–n* junction

This is the simplest junction available and is considered to consist of two grown regions of semiconductor, a *p*-type semiconductor doped with N_A acceptors per m^3, and a *n*-type semiconductor doped with N_D donors per m^3. The change from *p*-type to *n*-type is abrupt as indicated schematically in figure 10.11a. In equilibrium, the Fermi level or electrochemical potential in the two materials must be constant and

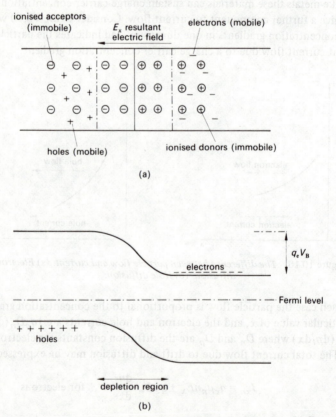

(a)

(b)

Figure 10.11   *The abrupt p–n junction.* (a) *Space charge region.* (b) *Energy band diagram.*

this leads to the valence and conduction bands bending as illustrated in figure 10.11b. The Fermi level in a *p*-type semiconductor is near the valence band and in a *n*-type semiconductor is near the conduction band (Allison, 1971). We shall deal only with a one-dimensional case.

10.7.1   The voltage barrier

The change in the electrostatic potential between the two materials discourages minority charge carrier injection, that is electron flow from n to p and hole flow p to n. In equilibrium, the electron and hole current flow is zero, and so, if E_x is the internal field existing in the transition region, then for electron flow

$$0 = q_e n \mu_n E_x + q_e D_n \frac{dn}{dx}$$

The diffusion constant is related to the mobility by the Einstein relationship $D_n = (kT/q_e)\mu_n$, thus

$$-q_e E_x \frac{dx}{kT} = \frac{dn}{n}$$

We assume that the transition region ends abruptly, so that $E_x = 0$ for $x > x_n$ or $x < -x_p$; this is called the *depletion approximation*, and the transition region is known as the *depletion region* since it is depleted of mobile charge-carriers. Now integrating over the transition region, if n_n is the density of electrons on the n side and n_p the density of electrons on the p side, then

$$\frac{n_n}{n_p} = \exp\left(\frac{q_e V_B}{kT}\right)$$

where V_B is the voltage barrier.

$$V_B = \frac{kT}{q_e} \ln\left(\frac{n_n}{n_p}\right) = \frac{kT}{q_e} \ln\left(\frac{p_p}{p_n}\right)$$

where p_p and p_n are the hole densities in the p and n material. This is known as the barrier voltage, built-in voltage or diffusion voltage.

10.7.2   The depletion region

If the assumption is made that all the donor and acceptor atoms are ionised then the net charge density $\rho = 0$ for $x > x_n$ and $x < -x_p$. In the depletion region this is not so and, assuming there are no mobile changes in the depletion region, Poisson's equation for the n-type material is

$$\frac{\partial^2 V}{\partial x^2} = -\frac{q_e N_D}{\epsilon_0 \epsilon_r} \qquad 0 < x < x_n$$

The boundary conditions for this region are

$$E_x = -\frac{\partial V}{\partial x} = 0 \qquad \text{at} \qquad x = x_n$$

and $V = 0$ at $x = 0$. For the p side

$$\frac{\partial^2 V}{\partial x^2} = \frac{+ q_e N_A}{\epsilon_0 \epsilon_r} \qquad -x_p < x < 0$$

with boundary conditions $E_x = 0$ at $x = -x_p$ and at $x = 0$.

Integrating we have

n side

p side

$$\frac{\partial V}{\partial x} = \frac{- q_e N_D}{\epsilon_0 \epsilon_r} (x - x_n) \qquad \frac{\partial V}{\partial x} = \frac{+ q_e N_A}{\epsilon_0 \epsilon_r} (x + x_p)$$

$$V = \frac{-q_e N_D}{2\epsilon_0 \epsilon_r} x(x - 2x_n) \qquad V = \frac{+q_e N_A}{2\epsilon_0 \epsilon_r} (x + 2x_p)$$

Thus for $\partial V / \partial x$ to equal 0 at $x = 0$

$$q_e N_D x_n = q_e N_A x_p$$

or the total positive space charge equals the total negative space charge. Assuming $V = - V_p$ at $x = -x_p$ and $V = V_n$ at $x = x_n$ then

$$V_B = V_n + V_p$$

$$= \frac{q_e}{2\epsilon_0 \epsilon_r} (N_A x_p{}^2 + N_D x_n{}^2)$$

Hence

$$x_p = \left[\frac{2\epsilon_0 \epsilon_r V_B N_D}{q_e N_A (N_A + N_D)} \right]^{1/2}$$

$$x_n = \left[\frac{2\epsilon_0 \epsilon_r V_B N_A}{q_e N_D (N_A + N_D)} \right]^{1/2}$$

If the total width of the depletion region is d, then

$$d = \left[\frac{2\epsilon_0 \epsilon_r V_B}{q_e} \frac{(N_A + N_D)}{N_A N_D} \right]^{1/2}$$

The relationships of ρ, E_x and $V(x)$ are plotted in figure 10.12.

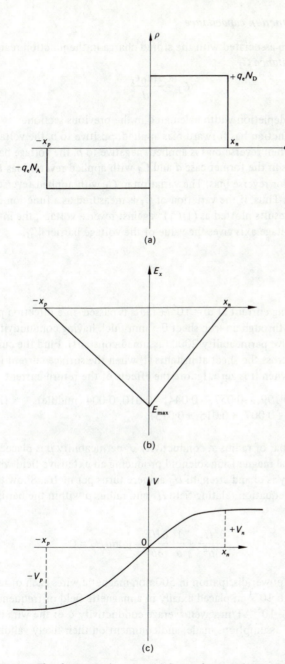

Figure 10.12   *The abrupt p–n junction. (a) Space charge variation. (b) Variation of electric field. (c) Variation of voltage.*

10.7.3 The junction capacitance

The capacitance associated with the stored charge in the junction region is the *junction capacitance* C_J

$$C_J = \frac{A\epsilon_0\epsilon_r}{d}$$

where d is the depletion width calculated in the previous section.

When the junction has forward bias applied, positive to p, the voltage barrier is reduced and, when reverse bias is applied, negative to p, the voltage barrier is increased. Thus in the former case d and C_J with applied reverse bias V_r is and vice versa for reverse bias. The variation in C_J with applied reverse bias V_r is $C_J \propto (1/V_r)^{1/2}$. Thus, if the variation of C_J is measured as a function of reverse bias, then the results plotted as $(1/C_J)^2$ against reverse voltage, the intercept on the negative voltage axis gives the value of the voltage barrier V_B.

Problems

10.1 Alternating current at $\omega = 10^5/\pi$ rad/s is passed, in a direction parallel to the surfaces, through an iron sheet 0·5 mm thick having conductivity $\sigma = 10^7$ S/m and relative permeability 2000 (assumed constant). Find the current distribution across the sheet at instants (i) when the surface current is a maximum and (ii) when it is zero. Ignore the effects of the return current.

((i) 1, 0·199, −0·057, −0·044, −0·010, 0·004 (middle);   (ii) 0, −0·310, −0·123, −0·007, +0·015, +0·013)

10.2 A metal bar of radius a, conductivity σ, permeability μ is placed axially in a cylindrical magnetising solenoid producing an extensive field whose angular frequency is ω and strength H_a ampere turns per metre. Show that the differential equation relating field H_z and radius ρ within the bar is the Bessel equation

$$\frac{\partial^2 H_z}{\partial\rho^2} + \frac{1}{\rho}\frac{\partial H_z}{\partial\rho} - j\omega\sigma\mu H_z = 0$$

Find the power dissipation in 500 non-magnetic wires each of radius 10^{-4} m and length 10^{-1} m placed axially in a magnetic field of frequency 10^6 Hz and amplitude 10^5 AT/m, given average conductivity σ of the wire is 10^7 S/m. State the assumptions made, and comment on their likely validity.

(6·1 kW)

10.3 A very long, thick, horizontal iron plate has, across its breadth b, two parallel slots of width s, depth d; the pitch between slot centre lines is p. The sides of a coil of N turns fill the bottoms of the slots to a depth c where $c < d \ll p$.

(a) Assuming the iron permeability to be very great, ignoring the effect of the coil 'ends' and assuming all the air flux to be horizontal, show that the self-inductance of the coil is approximately $\mu_0 N^2 b(2d - 4c/3)/s$ H.

(b) Another thick iron plate is placed over the first, with an air-gap clearance g; assuming all the air flux now to be vertical, show that the self-inductance of the coil is approximately

$$\mu_0 N^2 b \left[\frac{p - s}{2g} + \frac{s}{6(g + d)} \right] \text{ H}$$

(c) Comment on these two inductance 'components' given: b, s, d, p, c, g are 170, 2·5, 10, 110, 6, 2 cm respectively.

('Gap' component is 5·6 times 'slot' component)

10.4 A hollow glass cylinder of inner, outer radii ρ_1, ρ_2 has length $l(\gg \rho_2)$. It is uniformly wound *lengthwise* (that is as a long toroid) with N turns of wire carrying a current increasing uniformly with time $\partial I/\partial t$.

   If a particle, charged $+q$, is released inside a thin evacuated glass tube on the axis of the wound cylinder, find an expression for the energy it gains in travelling from one end to the other. Locate the correct cylinder end for entry.

   Evaluate this for $\rho_1, \rho_2, l, N, \partial I/\partial t, q = 2$ cm, 4 cm, 1 m, 100 turns, 10^8 A/s, 1 coulomb.

(1386 J/C)

10.5 A separately excited d.c. conduction-type electromagnetic pump is needed to deliver 0·05 m^3/s of liquid sodium at 5 m/s against a pressure of 2×10^5 N/m^2. The pole-face dimensions are 20 cm by 20 cm, the available air gap is 10 cm between faces and the flux density is 1 tesla.

   Find the active depth of liquid, the voltage and current needed at the electrodes and the overall efficiency. Assume active width of current-bearing sodium 20 cm; tube wall thickness 2 mm; resistivity of sodium at working temperature $0·18 \mu\Omega$m; resistivity of tube material $0·90 \mu\Omega$m; neglect excitation power loss and the 'leakage' resistance for material not in the active region.

(5 cm, 1·036 V, 14·8 kA, 65·5 per cent)

10.6 Show that, for an abrupt junction where the acceptor density in the p region is much greater than the donor density in the n region, the depletion width at zero bias is

$$\left[\frac{2\epsilon_0 \epsilon_r \mu_n V_B}{\sigma_n} \right]^{1/2}$$

where V_B is the voltage barrier, μ_n is the electron mobility and σ_n the electrical conductivity.

10.7 Poisson's equation for a semiconductor may be written

$$\frac{d^2 V}{dx^2} = - \frac{q_e(N + \bar{p} - \bar{n})}{\epsilon_0 \epsilon_r}$$

where N is positive for donors and negative for acceptors and $\bar{n}$ and $\bar{p}$ are the charge-carrier concentrations in thermal equilibrium of electrons and holes respectively. By considering drift and diffusion currents in thermal equilibrium show that

$$\frac{d^2 V}{dx^2} = \frac{-q_e}{\epsilon_0 \epsilon_r} \left[N - 2n_i \sinh\left(\frac{q_e V}{kT}\right) \right]$$

where the intrinsic charge-carrier density n_i is given by $n_i^2 = \bar{p}\,\bar{n}$.

10.8 An abrupt silicon junction has $N_A = 10^{15}/\text{cm}^3$, $N_D = 2 \times 10^{17}/\text{cm}^3$, $n_i = 10^{10}/\text{cm}^3$ and $\epsilon_r = 12$. Evaluate (i) the barrier voltage at 300 K, (ii) the width of the depletion region, (iii) the maximum electric field. Repeat (ii) and (iii) for a reverse voltage of 10 V.

((i) 0·71 V;   (ii) $\sim$9·5 x 10^{-5} cm;   (iii) $\sim$1·5 x 10^4 V/cm;
(ii) $\sim$3·6 x 10^{-4} cm;   (iii) $\sim$5·5 x 10^4 V/cm)

10.9 An alloyed p-n junction in germanium is made from p-type material with 10^{18} acceptors/cm^3 and n-type material with 8×10^{15} donors/cm^3, and is at 300 K. If the dielectric constant of germanium is 16 and n_i the intrinsic charge-carrier density is $2 \cdot 5 \times 10^{13}$ cm^{-3}, calculate (i) the voltage barrier, (ii) the depletion width.

((i) 0·43 V;   (ii) 3·1 x 10^{-7} m)

10.10 Show that depletion capacitance of a linearly graded p-n junction in which the impurity content varies as

$$|N_D - N_A| = ax$$

is

$$C = \left(\frac{\epsilon_0^2 \epsilon_r^2 q_e a}{12V} \right)^{1/3}$$

where V is the total voltage across the junction. Show that the maximum electric field strength in the junction (at $x = 0$) is given by

$$E_{\text{max}} = \frac{1}{2} \left(\frac{9}{4} \frac{q_e a V^2}{\epsilon_0 \epsilon_r} \right)^{1/3}$$

11

High-frequency Effects

In chapter 7 we studied the propagation of uniform plane electromagnetic waves in
free space and in media having loss. Although uniform plane waves are by definition
only an approximation, their use enables many electromagnetic wave problems to
be simplified, particularly in the field of optics and radio waves at large distances
from radiating antennae. Here we shall consider the propagation of uniform plane
waves across an interface and develop the elementary laws of reflection and refraction. The idea of reflection of waves from a good conductor will enable an introduction to be given to guided waves, with applications in microwave propagation.

11.1 Phase and group velocity

The wave equations for the electric and magnetic fields for non-conducting media
are

$$\nabla^2 \bar{E} = -\omega^2 \mu \epsilon \bar{E}$$

and

$$\nabla^2 \bar{H} = -\omega^2 \mu \epsilon \bar{H}$$

We have considered the general solution for these equations to be of the form

$$E = A \exp[j(\omega t - \beta z)] = A \exp\left[j\omega\left(t - \frac{z}{v_p}\right)\right]$$

The propagation constant, or phase constant, $\beta = \omega(\mu\epsilon)^{1/2}$ enables the *rate of
movement of a point of constant phase* to be calculated. The equation of motion
for the zero phase point is

$$\omega t - \beta z = 0$$

219

so

$$z = \frac{\omega}{\beta} t = v_p t$$

Here v_p is called the *phase velocity*, and is the velocity of propagation of a point of constant phase. The wave number or propagation constant has uses in other fields involving wave motion. In particular the wave number is an important concept in solid state studies where it is used to describe the motion of an electron in a periodic structure. Of particular practical interest are cases when the phase velocity is a function of frequency. This occurs in light waves in a medium in which the index of refraction varies with frequency, elastic waves in a medium consisting of discrete atoms bound by Hooke's law forces and electromagnetic-wave propagation in transmission lines with losses or in waveguides. It is of obvious importance when we consider a group of waves compounded from a set of waves each of different frequencies, for an observer in motion with one of the several wave velocities would see a constantly changing group. In this case, the concept of a *group velocity* is important. We can define the group velocity of a wave packet by considering a very simple group of waves made up of two waves of slightly differing frequencies, described by

$$A \sin[(\omega_0 - d\omega)t - (\beta_0 - d\beta)z]$$

and

$$A \sin[(\omega_0 + d\omega)t - (\beta_0 + d\beta)z]$$

The resultant of the two disturbances is

$$2A \sin(\omega_0 t - \beta_0 z)\cos(t\, d\omega - z\, d\beta)$$

This is a wave of angular frequency ω and a propagation constant β modified by a sinusoid of wavelength $2\pi/d\beta$ as shown in figure 11.1. Such a waveform will be

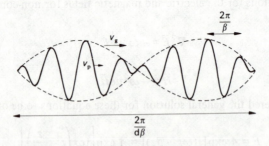

Figure 11.1   *Illustration of the formation of a wave group and the difference between phase and group velocity.*

familiar acoustically as an envelope of beats, a high frequency ω_0 whose amplitude varies at a much lower frequency $d\omega$. The envelope of waves moves with a velocity v_g which is the group velocity. The equation of motion of the zero phase point of

the envelope is

$$t\, d\omega - z\, d\beta = 0$$

Thus

$$z = \frac{d\omega}{d\beta}\, t = v_g t$$

and

$$v_g = \frac{d\omega}{d\beta}$$

For elastic waves in a homogeneous medium or for transverse electromagnetic waves in a vacuum $v_g = v_p$. When the phase velocity is a function of frequency the medium is said to be dispersive. If $dv_p/d\omega < 0$ the medium exhibits normal dispersion and when $dv_p/d\omega > 0$ it exhibits *anomalous dispersion*. A typical dispersion curve is shown in figure 11.2. Here two angles θ and ϕ are defined by $\tan \theta = \omega/v_p$ and $\tan \phi = dv_p/d\omega$. When $\tan \theta = \tan \phi$ which occurs at ϕ_c the group velocity is

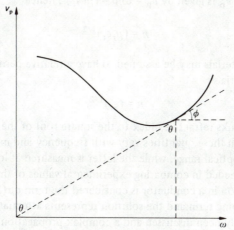

Figure 11.2 *Dispersion curve relationship between phase velocity v_p and angular frequency ω.*

infinite and for the particular curve shown the group velocity is negative for values of $\omega > \omega_c$. Should the medium have much dispersion then the compounded wave packet will change its shape rapidly while in motion so that the term group velocity has little meaning. Whenever energy flow takes place the velocity at which the energy is transferred is the group velocity.

11.2 The refractive index

The propagation of plane electromagnetic waves in dielectric and conducting material has been discussed previously. The equation representing the variation of electric field strength in a homogeneous and isotropic medium in which there is no free charge is

$$\nabla^2 E - \mu\epsilon\frac{\partial^2 E}{\partial t^2} - \sigma\mu\frac{\partial E}{\partial t} = 0$$

or in phasor form

$$\nabla^2\bar{E} + (\omega^2\epsilon - j\omega\mu\sigma)\bar{E} = 0$$

In a perfect dielectric the value of the conductivity, σ, is zero, there is no attenuation and the sinusoidal time variation leads to a space variation which is also sinusoidal. Now the *refractive index* of a material, n, is defined as the ratio of the velocity of electromagnetic radiation in free space (vacuum) to the velocity of radiation in the medium, that is

$$n = \frac{u_0}{v_p}$$

The phase velocity v_p is given by $v_p = \omega/\beta = (\mu\epsilon)^{1/2}$, hence

$$n = (\mu_r\epsilon_r)^{1/2}$$

Most dielectric materials may be assumed to have a relative permeability of $\mu_r = 1$ and for these materials

$$n = (\epsilon_r)^{1/2}$$

This relationship links refractive index to the square root of the dielectric constant, but in practice both these quantities vary with frequency and measurements of the former are in the optical range, while the latter is measured at lower frequencies. Care is therefore needed in comparing experimental values of these quantities.

When propagation in a conductor is considered the term $\sigma\mu(\partial E/\partial t)$ may be regarded as a damping term and the solution represents attenuated waves. This situation has already been discussed and a complex propagation constant γ is used to describe the wave

$$\gamma = [j\omega\mu(\sigma + j\omega\epsilon)]^{1/2}$$

The exponential propagation factor for phasor waves is often written as $\exp(-jkz)$ so that $jk = \gamma$, and the wave number k is obtained from

$$k^2 = \omega^2\epsilon\mu - j\omega\sigma\mu$$

An analogy with the case of propagation in a perfect dielectric medium may be made, and the equations representing the electromagnetic equations take the same form, if a complex dielectric constant

$$\epsilon_r = \epsilon_r' - j\epsilon_r''$$

is now introduced. Of course it is also possible to define a complex phase velocity and a complex refractive index such that

$$n = n' - jn''$$

In each case, the suffixes $'$ and $''$ refer to the real and imaginary parts.

The solution of the wave equation for the electric field strength may be re-written incorporating the real and imaginary parts of the refractive index so that

$$E = E_0 \exp\left(-\frac{\omega n'' z}{u_0}\right) \exp\left[j\omega\left(t - \frac{zn'}{u_0}\right)\right]$$

This is a wave of angular frequency ω and phase velocity u_0/n' with an amplitude modulation which is an exponential decay with distance. In section 7.5 propagation in a medium having loss was considered and the magnetic field strength was seen to have the same form but with a $\pi/4$ lag in phase. The power flow, obtained from the complex Poynting theorem, contains an exponentially decaying term, that is, $\exp(-2\omega n'' z/u_0)$.

The *absorption constant or coefficient* α describes the decay of intensity of a light wave on transmission through a material. The intensity at z, I_z, is related to the intensity I_0 at $z = 0$ by the relationship

$$I_z = I_0 \exp(-\alpha z)$$

It is possible therefore to relate the absorption coefficient to the imaginary part of the refractive index as follows

$$\alpha = \frac{2\omega n''}{u_0} = \frac{4\pi n''}{\lambda_0}$$

where λ_0 is the free-space wavelength. The imaginary part of the refractive index is often called the *absorption index*.

The wave number k may be related to the complex permittivity

$$k = \omega[\mu(\epsilon' - j\epsilon'')]^{1/2}$$

$$= \omega\left[\mu\epsilon'\left(1 - \frac{j\epsilon''}{\epsilon'}\right)\right]^{1/2}$$

and the ratio ϵ''/ϵ' is termed the *loss tangent*.

In many materials the imaginary part of the dielectric constant is negligible for frequencies below the microwave range and it is usual to describe the phase velocity in terms of the permittivity and permeability. This applies only to materials which have linear magnetic behaviour and any discussion of wave propagation in ferro-magnetic and ferrimagnetic materials would be out of place in this book. At fre-quencies above the microwave range, that is in the infrared and optical frequencies, it is usual to use the refractive index to characterise the properties of dielectrics.

11.3   Transmission and reflection of waves

Here we consider a wave, propagating in the z direction, incident on the interface
of two materials of refractive index n_1 and n_2. For convenience the interface is
taken as the xy plane and the electric field in the uniform plane wave is
assumed to be entirely in the x direction and of the general form

$$E_x = E_+ \exp[j(\omega t - \beta z)] + E_- \exp[j(\omega t + \beta z)]$$

as usual the $+$ and $-$ sign referring to propagation along the positive and negative
directions of z.

11.3.1   Waves incident upon a perfect conductor

The first case we shall consider will be for normal incidence. So that the boundary
condition for the electric field strength E is satisfied there must be a reflected wave
which when compounded with the incident wave gives zero electric field strength
at the interface. For a perfect conductor Poynting's theorem shows that there can
be no transmission and all the incident energy returns as a reflected wave. The
boundary condition is $E_x = 0$ at $z = 0$ for all values of time, thus $E_+ = E_-$ and
the expressions for the electric and magnetic fields are

$$E_x = E_+ [\exp(j\beta z) - \exp(j\beta z)] \exp(j\omega t)$$

$$= -2jE_+ \sin(\beta z) \exp(j\omega t)$$

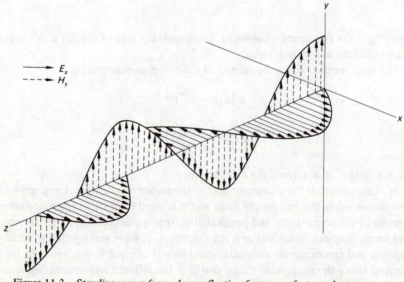

Figure 11.3   *Standing waves formed on reflection from a perfect conductor.*

and

$$H_y = \frac{2E_+}{Z} \cos(\beta z) \exp(j\omega t)$$

The total electric and magnetic fields are in time quadrature but are still perpendicular with their magnitudes related by the characteristic impedance of the incident region. This is an example of a *standing wave* since the electric field strength is zero at $z = 0$, the conductor surface, and also at distances $m\lambda/2$ in the negative z direction, where m is an integer. At these distances the magnetic field strength is a maximum value, and a minimum in the magnetic field strength and maximum in the electric field strength occurs at $z = -(2m + 1)\lambda/4$. This is illustrated in figure 11.3. In such a wave the average value of the Poynting vector at any cross-sectional plane is zero so that, as expected, as much energy is transported away from the surface by the reflected wave as is brought towards the surface by the incident wave. Twice each cycle all the energy is stored in the electric field and $\pi/2$ radians later it is all stored in the magnetic field.

11.3.2 Waves incident at an angle to a perfect conductor

When non-normal incidence takes place the two cases of interest which are convenient to illustrate are those with the electric field polarised in, and at right angles to, the plane of incidence. The plane of incidence contains the normal to the surface upon which the wave is incident and the direction of propagation of the wave. The two cases are illustrated in figure 11.4. Following the convention used in optics, the

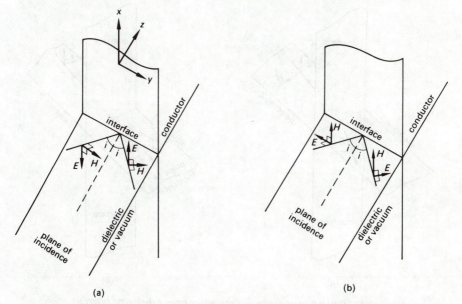

Figure 11.4 *Reflection from a conducting boundary. (a) Electric field normal to the plane of incidence. (b) Electric field in the plane of incidence.*

direction of propagation is described by a ray and in the incident and reflected wave the electric and magnetic field will be perpendicular to it. For a perfect conductor there will be no propagation past the surface of the conductor but in reality a weak highly attenuated wave is transmitted. In each case the transmitted ray is along the z direction with the electric and magnetic fields in the x and y directions. The particular direction for each field in the transmitted wave is determined by the initial polarisation associated with the incident wave. The transmitted wave will have its electric field leading its magnetic field by $\pi/4$.

11.3.3 Reflection and refraction of uniform plane waves at a dielectric interface

The interface between two media of refractive index n_1 and n_2 respectively is shown in figure 11.5. The media are non-conducting and the interface is in the xy plane, $z = 0$. The incident ray has been chosen to be in the $y = 0$ plane so that the x axis is in the plane of incidence and the incident ray is at an angle i to the z direction. We assume that the variations in field components of the incident wave are of the form

$$\exp\left\{j\omega\left[t - \frac{n_1(x \sin i + z \cos i)}{u_0}\right]\right\}$$

The boundary conditions for the electric and magnetic fields are by now familiar and there will be refracted and reflected waves obeying these conditions. The continuity of the tangential components of the field vectors is for all values of x and y

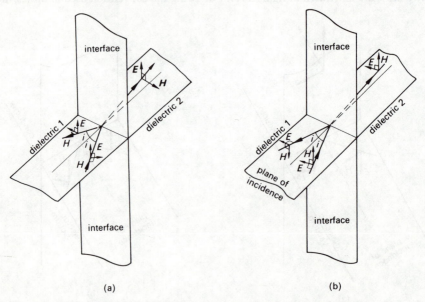

(a)   (b)

Figure 11.5 *Reflection of uniform plane waves from a dielectric interface.*
(a) *Electric field normal to plane of incidence.* (b) *Magnetic field normal to plane of incidence.*

and for all values of time. Here it is easy to see that the incident, reflected and refracted rays must all be in the same plane and the frequency of the waves must also be the same. Figure 11.5 also shows the direction of the polarisation and the angles i' and r are the angle of reflection and angle of refraction respectively.

Case 1 Electric field normal to the plane of incidence This is illustrated in figure 11.5a and the electric and magnetic field strengths in each wave are given.

For the incident wave

$$E_{yi} = E_{0i} \exp\left\{j\omega\left[t - \frac{n_1(x \sin i + z \cos i)}{u_0}\right]\right\}$$

$$H_{xi} = -\frac{E_{0i}}{Z_1} \cos i \exp\left\{j\omega\left[t - \frac{n_1(x \sin i + z \cos i)}{u_0}\right]\right\}$$

$$H_{zi} = \frac{E_{0i}}{Z_1} \sin i \exp\left\{j\omega\left[t - \frac{n_1(x \sin i + z \cos i)}{u_0}\right]\right\}$$

For the reflected wave

$$E_{yr} = E_{0r} \exp\left\{j\omega\left[t - \frac{n_1(x \sin i' - z \cos i')}{u_0}\right]\right\}$$

$$H_{xr} = +\frac{E_{0r}}{Z_1} \cos i' \exp\left\{j\omega\left[t - \frac{n_1(x \sin i' - z \cos i')}{u_0}\right]\right\}$$

$$H_{zr} = \frac{E_{0r}}{Z_1} \sin i' \exp\left\{j\omega\left[t - \frac{n_1(x \sin i' - z \cos i')}{u_0}\right]\right\}$$

For the transmitted wave

$$E_{yt} = E_{0t} \exp\left\{j\omega\left[t - \frac{n_2(x \sin r - z \cos r)}{u_0}\right]\right\}$$

$$H_{xt} = -\frac{E_{0t}}{Z_2} \cos r \exp\left\{j\omega\left[t - \frac{n_2(x \sin r - z \cos r)}{u_0}\right]\right\}$$

$$H_{zt} = \frac{E_{0t}}{Z_2} \sin r \exp\left\{j\omega\left[t - \frac{n_2(x \sin r - z \cos r)}{u_0}\right]\right\}$$

For the boundary conditions to be satisfied the exponential terms must be equal at the interface and

$$n_1 x \sin i = n_1 x \sin i' = n_2 x \sin r$$

Thus the elementary laws of optics have been predicted, namely that the angle of incidence equals the angle of reflection, and Snell's law, that is

$$n_1 \sin i = n_2 \sin r$$

Since the tangential components of the electric and magnetic field must be continuous through the interface

$$E_{0i} + E_{0r} = E_{0t}$$

and

$$\frac{(E_{0i} - E_{0r}) \cos i}{Z_1} = \frac{E_{0t} \cos r}{Z_2}$$

The relationship between the incident wave and the reflected and refracted waves may be expressed as follows.

$$\frac{E_{0r}}{E_{0i}} = \frac{Z_2 \cos i - Z_1 \cos r}{Z_2 \cos i + Z_1 \cos r}$$

$$\frac{E_{0t}}{E_{0i}} = \frac{2Z_2 \cos i}{Z_2 \cos i + Z_1 \cos r}$$

Case 2 Electric field within the plane of incidence   Now there are two components of electric field strength and only one component of magnetic field strength and the equations for the components of the incident wave are, see figure 11.5b

$$E_x = E_{0i} \cos i \exp\left\{j\omega\left[t - \frac{n_1(x \sin i + z \cos i)}{u_0}\right]\right\}$$

$$E_z = -E_{0i} \sin i \exp\left\{j\omega\left[t - \frac{n_1(x \sin i + z \cos i)}{u_0}\right]\right\}$$

$$H_y = \frac{E_{0i}}{Z_1} \exp\left\{j\omega\left[t - \frac{n_1(x \sin i + z \cos i)}{u_0}\right]\right\}$$

Proceeding as before we designate E_{0r} and E_{0t} to represent reflected and transmitted amplitudes, and the boundary conditions now give

$$(E_{0i} + E_{0r}) \cos i = E_{0t} \cos r$$

and

$$\frac{E_{0t} - E_{0r}}{Z_1} = \frac{E_{0t}}{Z_2}$$

which in turn allow us to write

$$\frac{E_{0r}}{E_{0i}} = \frac{Z_2 \cos r - Z_1 \cos i}{Z_2 \cos r + Z_1 \cos i}$$

and

$$\frac{E_{0t}}{E_{0i}} = \frac{2Z_2 \cos i}{Z_2 \cos r + Z_1 \cos i}$$

11.3.4 Fresnel's equations and the Brewster angle

When the electromagnetic wave passes through the interface of two good dielectrics for which $\mu_r = 1$ in each, then we have for the two cases considered

	E normal to plane of incidence	E within the plane of incidence
$\dfrac{E_{0r}}{E_{0i}}$	$\dfrac{\sin(r-i)}{\sin(r+i)}$	$\dfrac{\sin 2r - \sin 2i}{\sin 2r + \sin 2i}$
$\dfrac{E_{0t}}{E_{0i}}$	$\dfrac{2 \sin r \cos i}{\sin(r+i)}$	$\dfrac{4 \sin r \cos i}{\sin 2r + \sin 2i}$

These equations are often referred to as *Fresnel's equations* and may be used for transmission over any interface including that between a dielectric and a conductor. When the electric field is normal to the plane of incidence there is no direction of incidence for which the reflected wave is zero since r is not usually equal to i. If, however, the electric field is in the plane of incidence then when $\sin 2i = \sin 2r$ there is no reflection. In this case $i + r = \pi/2$, and the reflected ray is normal to the incident ray. This occurs whenever $i = \tan^{-1}(n_2/n_1)$ and the angle is called the *Brewster or polarising angle* θ_B. If an unpolarised wave is reflected from an interface at an angle of incidence equal to the Brewster angle then the reflected wave will be polarised with the electric field normal to the plane of incidence.

The reflected wave has its electric field either in phase or π radians out of phase with the electric field in the incident wave according to the relative magnitudes of the refractive index and the values of angle of incidence and angle of reflection. These phase changes are given by

E normal to plane of incidence	E within the plane of incidence
$n_2 > n_1$	$n_2 > n_1 \quad i < \theta_B$
$E_y \quad \pi$	$E_x \quad \pi$
$H_x \quad 0$	$E_y \quad 0$
$H_z \quad \pi$	$H_y \quad 0$

11.3.5 Propagation through thin films

For normal incidence upon an interface between two media of refractive index n_1 and n_2 if $\mu_{r_1} = \mu_{r_2} = 1$ we have

$$\frac{E_{0r}}{E_{0i}} = \frac{Z_2 - Z_1}{Z_2 + Z_1} = \frac{n_1 - n_2}{n_1 + n_2} = r$$

and

$$\frac{E_{0t}}{E_{0i}} = \frac{2Z_2}{Z_2 + Z_1} = \frac{2n_1}{n_1 + n_2} = t$$

As defined above, r is the reflection coefficient and t is the transmission coefficient. When we are concerned with power reflection and transmission across the interface we use the average value of $\mathscr{P}$, Poynting's vector, for the respective waves. Hence the ratios of the average energy fluxes per unit time per unit area are

Reflection   Transmission

$$R = \left| \frac{\mathscr{P}_{r\ av}}{\mathscr{P}_{i\ av}} \right| = \frac{E_{0r}^2}{E_{0i}^2} \qquad\qquad T = \left| \frac{\mathscr{P}_{t\ av}}{\mathscr{P}_{i\ av}} \right| = \frac{n_2}{n_1} \frac{E_{0t}^2}{E_{0i}^2}$$

$$= \frac{(n_1 - n_2)^2}{(n_1 + n_2)^2} \qquad\qquad\qquad = \frac{4n_1 n_2}{(n_1 + n_2)^2}$$

As we have previously discussed in section 11.2, the refractive index is complex and T and R will indicate the phase change on transmission and reflection. For low absorption, however, the above expressions hold with the real part of the refractive index used. The conservation of energy is satisfied since $R + T = 1$. We may express the electric field, assumed to be in the x direction, for a wave incident on an interface in the xy plane, $z = 0$, as

$$E_x = E_{0i} \exp\left[j\omega\left(t - \frac{n_1 z}{u_0} \right) \right]$$

which may be written

$$E_x = E_{0i} \exp(j\omega t) \exp(-j\phi z)$$

where

$$\phi = \frac{\omega(n_1' - jn_1'')}{u_0}$$

Now we shall apply this to the reflection and transmission through a thin film when the incident beam is normal to the film. The multiple reflections which occur in such a thin film are indicated in figure 11.6. Assuming that the internal reflection coefficients at surfaces a and b are r_a^+ and r_b^- then the external reflection coefficients at these surfaces are $-r_a^+ - r_b^-$. The transmission coefficients are t_a^+ and t_b^+ in the z direction and t_a^- and t_b^- in the negative z direction.

From considerations of conservation of energy we have

$$(r_a^+)^2 + t_a^+ t_a^- = 1$$

$$(r_b^-)^2 + t_b^+ t_b^- = 1$$

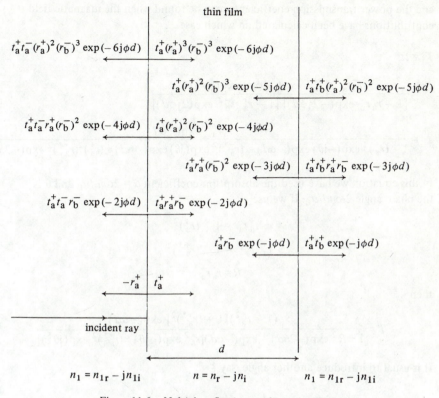

$$r = -r_a^+ + t_a^+ t_a^- r_b^- \exp(-2j\phi d) + t_a^+ t_a^- r_a^+ (r_b^-)^2 \exp(-4j\phi d)$$
$$+ t_a^+ t_a^- (r_a^+)^2 (r_b^-)^3 \exp(-6j\phi d) + \ldots$$
$$= -r_a^+ + t_a^+ t_a^- r_b^- [\exp(-2j\phi d)][1 - r_a^+ r_b^- \exp(-2j\phi d)]^{-1}$$

Figure 11.6 *Multiple reflections within a thin film.*

and by taking the sum of the contributions to the reflected and transmitted components we have

$$r = -r_a^+ + t_a^+ t_a^- r_b^- \exp(-2j\phi d) + t_a^+ t_a^- r_a^+ (r_b^-)^2 \exp(-4j\phi d)$$
$$+ t_a^+ t_a^- (r_a^+)^2 (r_b^-)^3 \exp(-6j\phi d) + \ldots$$
$$= -r_a^+ + t_a^+ t_a^- r_b^- [\exp(-2j\phi d)][1 - r_a^+ r_b^- \exp(-2j\phi d)]^{-1}$$

$$t = t_a^+ t_b^+ \exp(-j\phi d) + t_a^+ t_b^+ r_a^+ r_b^- \exp(-3j\phi d)$$
$$+ t_a^+ t_b^+ (r_a^+)^2 (r_b^-)^2 \exp(-5j\phi d) + \ldots$$
$$= t_a^+ t_b^+ [\exp(-j\phi d)][1 - r_a^+ r_b^- \exp(-2j\phi d)]^{-1}$$

If the medium is the same on either side of the film then

$$r_a^+ = r_b^-, \qquad t_a^+ = t_b^-, \qquad t_b^+ = t_a^-$$

and the power transmission coefficients may be found when the magnetic field contributions have been calculated, in which case

$$T = tt^*$$

$$= \frac{t_a^+ t_a^{+\,*} t_b^+ t_b^{+\,*} \exp(-j\phi d) \exp(j\phi^* d)}{[1 - r_a^+ r_b^- \exp(-2j\phi d)][1 - r_a^{+\,*} r_b^{-\,*} \exp(2j\phi d)]}$$

$$= \frac{t_a^+ t_a^{+\,*} \exp(-\alpha d)}{1 - (r_a^+)^2 \exp(-j\theta) \exp(-\alpha d) - (r_a^{+\,*})^2 \exp(j\theta) \exp(-\alpha d) + (r_a^+)^2 (r_a^{+\,*})^2 \exp(-2\alpha d)}$$

In this equation we have used the absorption coefficient $\alpha = 2\omega n_i/u_0$ and θ is the phase angle $2\omega n d/u_0$. If we use

$$t_a^+ t_a^- = 1 - (r_a^+)^2$$

$$t_a^{+\,*} t_a^{-\,*} = 1 - (r_a^{+\,*})^2$$

$$R = r_a^+ r_a^{+\,*}$$

then

$$T = \frac{(1 - r_a^{+2})[1 - (r_a^{+\,*})^2] \exp(-\alpha d)}{1 + R^2 \exp(-2\alpha d) - \exp(-\alpha d)[r_a^{+2} \exp(-j\theta) + (r_a^{+\,*})^2 \exp(j\theta^*)]}$$

It is usual to introduce another angle, say ξ

$$\xi = \tan^{-1}\left(\frac{2n'}{n''^2 + n'^2 + 1}\right)$$

giving

$$T = \frac{(1 - R)^2 + 4R \sin^2 \xi}{\exp(\alpha d) + R^2 \exp(-\alpha d) - 2R \cos(\theta + 2\xi)}$$

The $\cos(\theta + 2\xi)$ term will be familiar to students having a knowledge of optics since it is the term describing multiple-beam interference fringes.

11.4 Fibre Optics

Consider a wave transmitted into a dielectric 1 with refractive index n_1 which is surrounded by a second dielectric 2 with refractive index n_2, where $n_1 > n_2$; see figure 11.7.

If the wave is incident on n_2 at the critical angle, it will be *totally internally reflected* and an evanescent wave propagates along the boundary. (These aspects are discussed further in the following section.) Dielectric 1 becomes a waveguide and such waves may be propagated long distances with only small attenuation.

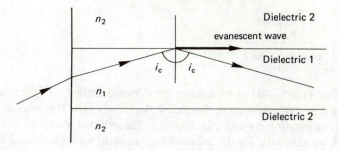

Figure 11.7 *Total internal reflection at the critical angle.*

In fibre optics the inner dielectric consists of a cylindrical core of glass or plastic which is covered by a cladding of lower refractive index material. The core and cladding are drawn in a similar manner to the drawing of copper wire, resulting in the production of a narrow filament which may be only a few micrometres in diameter.

11.4.1 Critical angle of incidence and the evanescent wave

The angel of refraction (r) in terms of the incident angle (i) is

$$\cos r = \left[1 - \left(\frac{n_1}{n_2} \sin i \right)^2 \right]^{1/2}$$

For a *critical angle* $i = i_c$, $(n_1/n_2) \sin i_c = 1$, and $\cos r = 0$ (that is $r = 90°$). At the critical angle the reflection coefficient (see section 11.3.5) is

$$R = \left(\frac{n_1 \cos i - n_2 \cos r}{n_1 \cos i + n_2 \cos r} \right)^2 = 1$$

and since $R + T = 1$, to satisfy energy conservation, then $T = 0$. Hence, for a wave incident at the critical angle, all the energy appears in the reflected wave, as shown in figure 11.7.

However, for E normal to the plane of incidence we have

$$\left. \frac{E_{0t}}{E_{0i}} \right|_{\mathrm{N}} = \frac{2z_2 \cos i}{z_2 \cos i + z_1 \cos r} = \frac{2n_1/n_2 \cos i}{\left(\dfrac{n_1}{n_2} \right) \cos i + \cos r}$$

$$= 2 \text{ when } i = i_c \text{ and } \cos r = 0$$

Similarly for E parallel to the plane of incidence

$$\frac{E_{ot}}{E_{oi}}\bigg|_P = \frac{2\left(\dfrac{1}{n_2}\right)\cos i}{\dfrac{1}{n_1}\cos i + \dfrac{1}{n_2}\cos t} = \frac{2n_1}{n_2}$$

Hence, the amplitude E_{ot} is not zero. Since the refracted angle is $90°$, then the transmitted wave travels through the boundary along the x axis. This wave is known as an *evanescent wave* and it is coupled to the reflected wave.

In general, we may write for the refracted (transmitted) wave (see page 227)

$$E_r = E_{or} \exp j \left[\omega t - k_2 \ (x \sin r - z \cos r)\right]$$

where $k_2 = \omega n_2 / u_0$. We may epxress this in terms of the incident angle i only using

$$\cos r = \sqrt{1 - \sin^2 r} = j \sqrt{\left(\frac{n_1}{n_2} \sin i\right)^2 - 1}$$

and $k_1 \sin i = k_2 \sin r$. Therefore

$$E_r = E_{ot} \exp j [\omega t - k_2 x \sin r] \ \exp (jk_2 z \cos r)$$

and

$$E_r = E_{ot} \exp j [\omega t - k_1 x \sin i] \ \exp - k_2 z \left[\left(\frac{n_1}{n_2} \sin i\right)^2 - 1\right]^{1/2}$$

If $(n_1/n_2) \sin i > 1$, then E_t decreases exponentially in the z direction.

If $(n_1/n_2) \sin i = 1$, the exp (z) terms $\rightarrow$ unity and the refracted wave only has an x component. This is the evanescent wave which travels along the boundary between the two media. The evanescent wave absorbs zero energy from the incident wave since $R = 1$ at the critical angle.

11.4.2 Propagation and losses

In figure 11.8 the angle θ_a is the maximum angle at which a wave incident on the coupling interface can enter the core for total internal reflection. For angles greater than θ_a such waves will be only partially reflected at the core-cladding interface and they decay rapidly along the fibre. The angle θ_a is known as the acceptance angle and $\sin \theta_a$ is the numerical aperture (NA). From Snell's law

$$n_0 \sin \theta_a = n_1 \sin (90 - i_c) = n_1 \cos i_c$$

$$= n_1 \sqrt{1 - \sin^2 i_c} = n_1 \sqrt{1 - \left(\frac{n_2}{n_1}\right)^2}$$

If $n_0 = 1$, that is air, the numerical aperture is

$$NA = \sin \theta_a = \sqrt{n_1^2 - n_2^2}$$

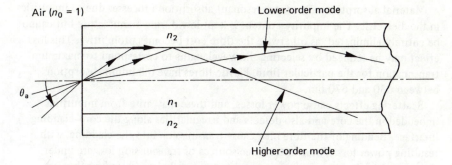

Figure 11.8 *Optical fibre, showing numerical aperture and propagation modes.*

The waves which enter the fibre are classified as *meridional* and *skew*. The former pass through the axis of the fibre, while the latter do not pass through the axis. In addition, there are parallel waves propagating along the core–cladding interface. These are the evanescent waves. The meridional waves consist of low-order modes which arise from waves incident at small angles within the acceptance angle and higher-order modes which enter at larger angles. A *single mode* fibre is one in which NA is so small that only one mode propagates.

11.4.3 Dispersion

There are two principal sources of dispersion in optical fibres. These are material dispersion and model dispersion. Material dispersion arises because different wavelengths travel at different velocities in solid dielectrics. Hence, if the signal source consists of a number of wavelengths, these will not arrive at the receiver simultaneously. Model dispersion arises owing to a difference in path lengths between low-order modes and higher-order modes. The higher-order modes have a longer travel time, so that the various modes will be time-dispersed on arrival at the receiver.

11.4.4 Transmission losses

The main transmission losses are

(1) Material absorption
(2) Scattering from defects and impurities
(3) Scattering due to defects at the interface between core and cladding
(4) Radiation losses due to bending of the fibre
(5) Coupling losses

Material absorption arises from resonant absorption processes due to impurities in the dielectric. The impurity content is kept low during manufacture but cannot be entirely eliminated, as otherwise the fibre cost becomes prohibitive. This loss effect may be reduced by selecting the wavelengths to correspond to maximum transmission for the particular fibre. Plastic fibres have minimum absorption between 630 and 670 nm.

Scattering effects cause power losses, and these may arise from impurities imbedded in the core and also defects and irregularities along the core–cladding interface. Bending of the fibre may cause transmission into the cladding, with resulting power loss. These latter three sources of transmission loss are under the control of the manufacturer and cannot normally be reduced by the user.

Coupling losses arise in the emitter to fibre connector, fibre to fibre connectors and fibre to detector connector. These losses are minimised by good coupling design, which attempts to reduce reflections at mating surfaces and achieve exact alignment between optical surfaces.

The total power losses in a fibre optic communications link may be estimated by summing all of the above losses. The individual losses (coupling, cable, etc.) are usually given in manufacturers' data sheets, and an example of such a calculation follows.

11.4.5 *Example of fibre optic loss calculations*

The coupling losses associated with a fibre optic link are specified as -1 dB *each* for the length of fibre and the transmitter and receiver connectors.

(a) If the power output from the transmitter is 25 μW, estimate the power input to the receiver. (b) If the receiver responsivity is 0.5 A/W, estimate the current into the receiver.

(a) The attenuation $\alpha = 10 \log\left(\dfrac{\text{Power out}}{\text{Power in}}\right)$ dB

In this case $\alpha = -3$ dB and $P_{in} = 25$ μW, so

$$\log P_{out} = \frac{1}{10}(-3 + 10 \log P_{in}) = -0.3 + \log 25 = 1.098$$

Therefore $P_{out} = 12.53$ μW. Hence, only half of the input power reaches the receiver.

(b) If the receiver responsivity is 0.5 A/W, the current into the receiver is $0.5 \times 12.53 = 6.26$ μA.

11.5 Electromagnetic waves on a transmission line

Transmission lines consist of any two conductors separated by a dielectric and may be coaxial cylindrical conductors, parallel wires or parallel plates. The conventional treatment of transmission lines is by means of distributed parameters, inductance, capacitance, resistance and conductance per unit length with equal and opposite currents flowing at a given plane through the conductors and a voltage developed between them. In the elementary approach encountered at an early stage in network courses, series resistance and shunt capacitance are neglected and the analysis is by means of a differential length of line dz and its associated series inductance $L\,dz$ and shunt capacitance $C\,dz$ as described in figure 11.9. The expression for

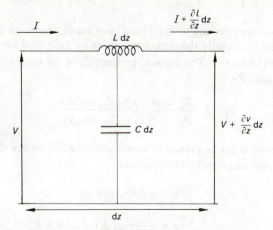

Figure 11.9 *Differential length of transmission line.*

current and voltage may be combined in the following two equations

$$\frac{\partial^2 V}{\partial z^2} = LC\frac{\partial^2 V}{\partial t^2}$$

$$\frac{\partial^2 I}{\partial z^2} = LC\frac{\partial^2 I}{\partial t^2}$$

The expression for voltage may be rewritten

$$\frac{\partial^2 V}{\partial z^2} = \frac{1}{u^2}\frac{\partial^2 V}{\partial t^2}$$

the usual form of the wave equation with $u = (LC)^{-1/2}$, and this has the general solution

$$v = f_1(ut - z) + f_2(ut + z)$$

A similar relationship holds for current and the ratio of voltage to current; the characteristic impedance Z_0 of the line, is

$$Z_0 = \left(\frac{L}{C}\right)^{1/2}$$

It is an easy matter to show that no reflection occurs from a matched discontinuity in the line which has an impedance equal to Z_0.

Most practical applications involve sinusoidal time variations leading to a voltage given by

$$V = \left[V_+ \exp\left(\frac{-j\omega z}{u} \right) + V_- \exp\left(\frac{j\omega z}{u} \right) \right] \exp(j\omega t)$$

with the $+$ and $-$ sign referring to the direction of travel. Here ω/u is referred to as the phase constant β, exactly the same notation as we have used for the electromagnetic wave in chapter 7. The input impedance Z_i of the simple line of length l, and load impedance Z_L is

$$Z_i = Z_0 \frac{Z_L \cos(\beta l) + j Z_0 \sin(\beta l)}{Z_0 \cos(\beta l) + j Z_L \sin(\beta l)}$$

and standing waves occur as a result of terminating the line with a short circuit. The voltage and current can now be represented as

$$V = -2j V_+ \sin(\beta z)$$

$$I = \frac{2V_+}{Z_0} \cos(\beta z)$$

which are obviously of the same form as that of the standing waves formed by reflection of a uniform plane wave from a perfectly conducting boundary. The same energy transfer from voltage to current as occurred from electric field to magnetic field is also present. The analogy between a uniform plane wave and a transmission line is given below.

Uniform plane wave

$$E_x(z) = E_+ \exp(-j\beta z) + E_- \exp(j\beta z)$$

$$H_y(z) = \frac{1}{Z_0} \{ E_+ \exp(-j\beta z) - E_- \exp(j\beta z) \}$$

$$\beta = \omega(\mu\epsilon)^{1/2}$$

$$Z_0 = \left(\frac{\mu}{\epsilon} \right)^{1/2}$$

Simple transmission line

$$V(z) = V_+ \exp(-j\beta z) + V_- \exp(j\beta z)$$

$$I(z) = \frac{1}{Z_0} \{ V_+ \exp(-j\beta z) - V_- \exp(j\beta z) \}$$

$$\beta = \omega(LC)^{1/2}$$

$$Z_0 = \left(\frac{L}{C}\right)^{1/2}$$

Further analogies are given in Ramo *et al.* (1965).

At higher frequencies, coaxial lines are invariably used in preference to parallel wires since they are easy to make with a dielectric used to support the inner conductor. Shielding is easy for coaxial lines but higher impedances may be achieved

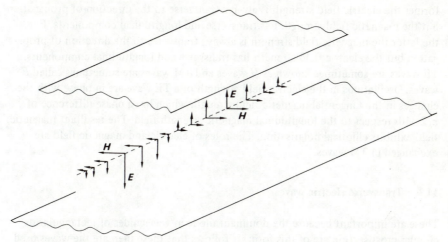

Figure 11.10 *Transverse electromagnetic wave between two infinite parallel-plate conductors.*

with the inherently balanced parallel wire transmission line. However, as an example of the type of wave appearing on a transmission line we consider the infinite parallel-plate pair shown in figure 11.10.

The sinusoidal currents in the conductors give rise to sinusoidal magnetic fields obeying the boundary conditions that the magnetic field strength *H* must be tangential. Since the voltage is in phase with the current this leads to a sinusoidal electric field with strength *E* normal to the conductors. This is a uniform linearly polarised transverse electromagnetic wave. For a coaxial line the field configuration would have a radial electric field strength and a concentric magnetic field strength. The variations of the magnetic and electric field correspond therefore to identical variations in current and voltage. Not surprisingly both approaches yield the same results.

When frequencies of operation do not exceed 3000 MHz or 3 GHz the transmission line is the standard device. The transmission is by means of a simple transverse electromagnetic wave with a wavelength at this frequency of 10 cm. When, however, the wavelength approaches the separation of the conductors, other waves occur and for waves with a wavelength less than 10 cm losses increase rapidly which prohibits the use of transmission lines. Instead, at these frequencies hollow conducting tubes are used to transfer power in the form of electromagnetic waves.

The waves are guided by the tube which is therefore not surprisingly called a *wave-guide*. We have previously discussed the nature of transverse electromagnetic waves in which the electromagnetic field strength vectors were mutually parallel. These waves can be guided by two, or more, separate conductors but because of the boundary conditions such waves cannot propagate along a hollow conducting tube. A TEM wave in an enclosed conductor would be a violation of the boundary conditions. The two types of wave which can exist within an enclosed conducting tube are the transverse electric (TE) and transverse magnetic (TM) waves. In the former the electric field strength is always transverse to the direction of propagation but the magnetic field strength has transverse and longitudinal components. For the latter the magnetic field strength is always transverse to the direction of propagation but the electric field strength has transverse and longitudinal components. TE waves are sometimes known as *H* waves and TM waves are sometimes called *E* waves. Oscillations in the electric field strength of a TE wave are in phase with the changes in the tangential magnetic field strength which has a phase difference of $\pi/2$ with respect to the longitudinal magnetic strength field. The resultant magnetic field exhibits elliptical polarisation. The roles of electric and magnetic field are exchanged in TM waves.

11.6   Transverse electric waves

These are important because the dominant mode in waveguides of rectangular or circular cross-section are of this form. It follows that these then are the waves used in microwave engineering.

11.6.1   Formation of TE waves

Let us first consider the incidence of a uniform plane wave upon a perfectly conducting plane and assume that the wave is polarised with its electric field strength vector parallel to the conducting surface. Following usual conventions the full line represents crests in the electric field strength with the resulting vector directed out of the plane of the page, while the dotted line represents wave troughs with the direction of the electric field strength into the plane of the paper. This is illustrated in figure 11.11. The wave is incident at an angle to the conductor which is considered to be perfect; thus the angle of incidence equals the angle of reflection and the electric field strength vector undergoes a change of phase on reflection. At A the maximum electric field points out of the plane of the paper, at B the field is a maximum in the opposite direction, while at C the field is zero. The usual boundary conditions hold, hence the electric field is zero along the conducting boundary and also along such planes as PP, QQ, etc. Since the incident wave is a uniform plane wave the magnetic field strength vector is in the plane of the page and is tangential at the boundary.

The superposition of the incident and reflected waves means that along planes such as RR there is a sinusoidal variation in the electric field strength and also along planes such as SS perpendicular to the conducting boundary. The resultant magnetic

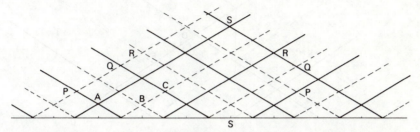

Figure 11.11   *Uniform plane wave incident upon a perfect conductor. The electric field is normal to the plane of the page.*

field configuration is a set of loops with the magnetic field in opposite direction in adjacent sets of loops. This change corresponds to the change in direction of the electric field.

As the uniform plane wave progresses in time the resultant pattern of electric and magnetic fields moves parallel to the conducting boundary. There is no movement of the pattern in a direction perpendicular to the plane of the page. The wave pattern is propagated with the electric field always transverse to the direction of propagation. Hence this is an example of a TE mode. We see how it it possible to propagate this mode in a waveguide since placing a conducting boundary along planes such as PP, QQ, perpendicular to the plane of the page and parallel to the original conducting boundary, does not affect the boundary conditions. These waves may then be propagated between two parallel conducting-plates. If two further plates were now added parallel to the plane of the page then the electric field would be normal and the magnetic field tangential to them. The boundary conditions are again satisfied. Hence it is possible to propagate a TE mode along a rectangular conducting pipe.

The wave we have described has a simple pattern and other more complicated patterns are possible in a rectangular guide which do not violate the boundary conditions. It is necessary therefore to define the patterns or modes by an easily recognisable nomenclature. This is done by considering the cross-section of the guide and calculating the number of half sinusoidal variations in electric field, first along the broad dimension, say m, and then n along the narrow dimension. These integers are then added as subscripts to the description of the wave. A TE_{11} mode would occur with one half sinusoidal variation of electric field along both directions. We shall see later that each mode has its own *cut-off wavelength* which is the maximum wavelength which can be propagated within the constraints of the tube.

11.6.2 TE_{mo} waves

Consider the diagram in figure 11.12 which depicts part of the pattern of the uniform plane wave reflected from the conductor. Conducting planes could be placed

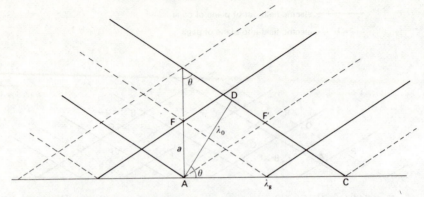

Figure 11.12   *Expanded portion of part of the uniform plane wave reflected from a conducting plane.*

at certain planes such as to confine the TE waves to a restricted region.

Here

$$AD = \lambda_0 \text{ the free-space wavelength}$$

$$AC = \lambda_g \text{ the wavelength in the guide}$$

and

$$AF = a$$

We have $\sin \theta = \lambda_0/2a$, $\cos \theta = \lambda_0/\lambda_g$ and

$$\left(\frac{\lambda_0}{\lambda_g}\right)^2 + \left(\frac{\lambda_0}{2a}\right)^2 = 1$$

Therefore

$$\frac{1}{\lambda_g^2} + \frac{1}{4a^2} = \frac{1}{\lambda_0^2}$$

Writing $\lambda_c = 2a$ we have

$$\frac{1}{\lambda_0^2} = \frac{1}{\lambda_g^2} + \frac{1}{\lambda_c^2}$$

λ_c is the cut-off wavelength and no wave is propagated with $\lambda_0 > \lambda_c$. The particular mode we have considered is a TE_{10} mode but the treatment is readily extended to TE_{mo} modes since for these $\lambda_c = 2a/m$. Propagation with $\lambda_0 > \lambda_c$ would give an imaginary λ_g; such a wave decays rapidly and is said to be *evanescent*. A typical waveguide used for 10 GHz wave propagation, waveguide 16, has internal dimensions of 0·9 in by 0·4 in. With the electric field strength E parallel to the narrow dimension $\lambda_c = 4.572$ cm for the TE_{10} mode, but for the same mode with E parallel to the broad dimension $\lambda_c = 2·032$ cm. At 10 GHz in the first case $\lambda_0 < \lambda_c$ but in the second case $\lambda_0 > \lambda_c$ and the wave is evanescent. With E parallel to the narrow dimension only the TE_{10} mode is propagated.

The wave equation in rectangular waveguide   In chapter 7 we considered applications of Maxwell's equations to electromagnetic wave propagation in free space and in a medium having loss. Now we wish to solve Maxwell's equations for propagation along a hollow metal pipe.

The wave equations for a non-conducting medium are

$$\nabla^2 E + \omega^2 \mu \epsilon E = 0$$

and

$$\nabla^2 H + \omega^2 \mu \epsilon H = 0$$

These equations may be applied to the rectangular guide to give solutions for the electric and magnetic field strength and by the use of the auxiliary relationships

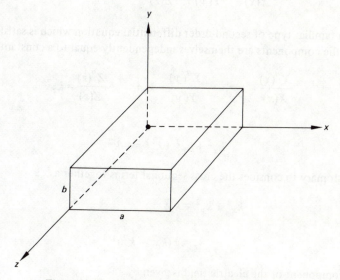

Figure 11.13   *Rectangular waveguide coordinates.*

those for magnetic and electric flux density. The problem is simplified by considering propagation along the z direction as indicated in figure 11.13. We first turn our attention to the z-directed components, namely

$$\frac{\partial^2 E_z}{\partial x^2} + \frac{\partial^2 E_z}{\partial y^2} + \frac{\partial^2 E_z}{\partial z^2} = -\omega^2 \mu \epsilon E_z$$

and

$$\frac{\partial^2 H_z}{\partial x^2} + \frac{\partial^2 H_z}{\partial y^2} + \frac{\partial^2 H_z}{\partial z^2} = -\omega^2 \mu \epsilon H_z$$

The solution of these equations is by means of separation of the variables and the solution for the electric field is assumed to be

$$E_z = X(x)Y(y)Z(z)$$

where $X(x)$, $Y(y)$ and $Z(z)$ are functions of x, y and z only, respectively, which gives

$$\frac{\partial^2 E_z}{\partial x^2} = X''(x)Y''(y)Z''(z)$$

and similar expressions for $(\partial^2 E_z/\partial y^2)$ and $(\partial^2 E_z/\partial z^2)$. These are now substituted into the z-component equation and the result divided by $X(x)Y(y)Z(z)$ to give

$$\frac{X''(x)}{X(x)} + \frac{Y''(y)}{Y(y)} + \frac{Z''(z)}{Z(z)} = -\omega^2 \mu \epsilon = -k^2$$

This is a familiar type of second-order differential equation which is satisfied only if all of the components are themselves independently equal to a constant, that is

$$\frac{X''(x)}{X(x)} = k_x{}^2, \qquad \frac{Y''(y)}{Y(y)} = k_y{}^2, \qquad \frac{Z''(z)}{Z(z)} = k_z{}^2$$

and

$$k_x{}^2 + k_y{}^2 + k_z{}^2 = -k^2$$

It is customary to consider the cross-sectional terms together as

$$k_x{}^2 + k_y{}^2 = k_c{}^2$$

so

$$k_z = \pm(k_c{}^2 - k^2)^{1/2}$$

The z component of the electric field is given by

$$\frac{\partial^2 E}{\partial z^2} = k_c{}^2 E_z$$

and the propagation is assumed to be without attenuation, so

$$E_z = E_0 \exp(\mathrm{j}\omega t - \gamma z)$$

with

$$\gamma^2 = -\beta^2$$

Thus

$$E_z = X(x)Y(y)\exp(\mathrm{j}\omega t - \beta z)E_0$$

The boundary conditions are

$$E_t = 0 \qquad \text{at } x = 0, x = a$$

$$E_t = 0 \qquad \text{at } y = 0, y = b$$

so suitable functions for the $X(x)$ and $Y(y)$ functions are trigonometrical and of the form $\sin(m\pi x/a)$, $\sin(n\pi y/b)$ respectively. Here m and n are integers or zero. The solution is therefore

$$E_z = E_0 \sin\left(\frac{m\pi x}{a}\right) \sin\left(\frac{m\pi y}{b}\right) \exp[j(\omega t - \beta z)]$$

Considering only the equation curl $E = -\dot{B} = -\mu\dot{H}$ and substituting for $\partial/\partial z = -j\beta$ we have

$$\frac{\partial E_z}{\partial y} + j\beta E_y = -j\omega\mu H_x$$

$$-j\beta E_x + \frac{\partial E_z}{\partial x} = -j\omega\mu H_y$$

$$\frac{\partial E_y}{\partial x} - \frac{\partial E_x}{\partial y} = -j\omega\mu H_z$$

The expressions for the x- and y-directed components in terms of the z-directed components derivatives may be arranged to give

$$\beta E_x - \omega\mu H_y = j\frac{\partial E_z}{\partial x}$$

$$\beta E_y + \omega\mu H_z = j\frac{\partial E_z}{\partial x}$$

If this procedure is now repeated for the curl H equation we arrive at two further equations, namely

$$\omega\epsilon E_x - \beta H_y = j\frac{\partial H_z}{\partial y}$$

$$\omega\epsilon E_y + \beta H_x = j\frac{\partial H_z}{\partial x}$$

These may be rewritten in terms of k_c to give

$$E_x = \frac{-j}{k_c^2}\left(\beta\frac{\partial E_z}{\partial x} + \omega\mu\frac{\partial H_z}{\partial y}\right)$$

$$E_y = \frac{j}{k_c^2}\left(-\beta\frac{\partial E_z}{\partial y} + \omega\mu\frac{\partial H_z}{\partial x}\right)$$

$$H_x = \frac{j}{k_c^2}\left(\omega\epsilon\frac{\partial E_z}{\partial y} - \beta\frac{\partial H_z}{\partial y}\right)$$

$$H_y = \frac{-j}{k_c^2}\left(\omega\epsilon\frac{\partial E_z}{\partial x} + \beta\frac{\partial H_z}{\partial y}\right)$$

In the above equations H_z and E_z are independent variables and two different types of solution are possible, namely those for which $E_z = 0$ and those for which $H_z = 0$. Thus we have TE and TM modes.

The transverse electric (TE) *mode*   In this $E_z = 0$ and if $E_y = 0$ then $\partial H_z/\partial x = 0$, and we have

similarly
$$\frac{\partial H_z}{\partial x} = 0 \qquad \text{at } x = 0 \text{ and } x = a$$

$$\frac{\partial H_z}{\partial y} = 0 \qquad \text{at } y = 0 \text{ and } y = b$$

The solution for the magnetic field variation is thus

$$H_z = H_0 \cos\left(\frac{m\pi x}{a}\right)\cos\left(\frac{n\pi y}{b}\right)\exp\{j(\omega t - \beta z)\}$$

This may now be used with the result of the previous section to give

$$E_x = \frac{j\omega\mu H_0}{k_c^2}\frac{n\pi}{b}\cos\left(\frac{m\pi x}{b}\right)\sin\left(\frac{n\pi y}{b}\right)$$

$$E_y = -\frac{j\omega\mu H_0}{k_c^2}\frac{m\pi}{a}\sin\left(\frac{m\pi x}{a}\right)\cos\left(\frac{n\pi y}{b}\right)$$

$$H_x = \frac{j\beta H_0}{k_c^2}\frac{m\pi}{a}\sin\left(\frac{m\pi x}{a}\right)\cos\left(\frac{n\pi y}{b}\right)$$

$$H_y = \frac{j\beta H_0}{k_c^2}\frac{n\pi}{b}\cos\left(\frac{m\pi x}{a}\right)\sin\left(\frac{n\pi y}{b}\right)$$

$$H_z = H_0 \cos\left(\frac{m\pi x}{a}\right)\cos\left(\frac{n\pi y}{b}\right)$$

The factor $\exp[j(\omega t - \beta z)]$ has to be included in all the above equations. For the TE$_{10}$ mode, the dominant mode in rectangular waveguide, $m = 1, n = 0$ and $\lambda_c = 2a$, hence

$$E_x = 0, \qquad E_z = 0$$

$$E_y = -j\frac{\lambda_c}{\lambda_0}\left(\frac{\mu}{\epsilon}\right)^{1/2}H_0 \sin\left(\frac{\pi x}{a}\right)\exp[j(\omega t - \beta z)]$$

$$H_x = j \frac{\lambda_c}{\lambda_g} H_0 \sin\left(\frac{\pi x}{a}\right) \exp[j(\omega t - \beta z)]$$

$$H_y = 0, \qquad H_z = H_0 \cos\left(\frac{\pi x}{a}\right) \exp[j(\omega t - \beta z)]$$

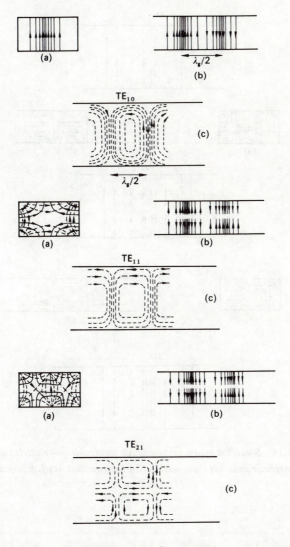

Figure 11.14 *Some* TE *modes in rectangular waveguide:* — *electric field;* − − − *magnetic field.* (a) *Cross-section.* (b) *Longitudinal.* (c) *Top surface.*

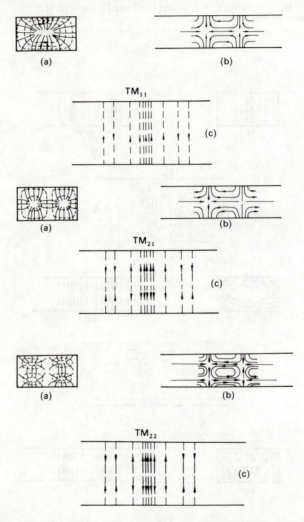

Figure 11.15   *Some* TM *modes in rectangular waveguide:* — *electric field;*
– – – magnetic field. (a) *Cross-section.* (b) *Longitudinal.* (c) *Top surface.*

This is the simplest electromagnetic wave pattern with the electric field in the y direction only and the magnetic field consisting of loops in the xz plane. Some TE mode patterns are shown in figure 11.14 and for comparison some TM patterns are given in figure 11.15.

Problems

11.1 The phase velocity v_p and group velocity v_g are ω/β and $d\omega/d\beta$ respectively. If λ is the corresponding wavelength, prove the two relationships

$$v_p = v_g + \lambda \frac{dv_p}{d\lambda}$$

$$v_p^{-1} = v_g^{-1} + \frac{\omega}{v_p^2} \frac{dv_p}{d\omega}$$

Referring to figure 11.2, show that the following relationship holds at γ

$$\frac{\omega}{v_p} \frac{dv_p}{d\omega} = 1$$

11.2 Consider an incident wave polarised with its electric vector normal to the plane of incidence impinging at an angle to the surface of an infinite conductor whose plane is perpendicular to the plane of incidence. Show that the transmitted wave has its magnetic vector parallel to the interface and that the transmitted wave is highly attenuated. Repeat the calculation for the incident wave polarised with the magnetic field within the plane of incidence.

11.3 From the definition of reflection and transmission coefficients, deduce that at the Brewster angle there is no component of electric field in the plane of incidence in the reflected wave.

For a wave incident at an angle to the normal, plot the coefficients of reflection and transmission, for $3n_1 = 2n_2$, as the angle of incidence varies from 0 to 90°. Show that for $n_1 = n_2$ the two coefficients are equal.

11.4 An electromagnetic wave polarised with its electric field normal to the plane of incidence passes from an optically denser to an optically rarer medium. Show that, as the angle of incidence increases, a value is reached at which total reflection occurs. Assume that the permeability of each medium is the same.

11.5 A plane polarised wave is incident normally upon a thin film, of thickness d, of conducting material. The incident wave has the electric field within the plane of incidence. Neglecting the multiple reflections occurring within the film, calculate, at the points A and B just inside each surface of the conducting film, (i) the electric and magnetic field, (ii) the average value of the Poynting vector.

If the film is of copper of conductivity 6×10^7 S/m, of skin depth 66 μm and of 100 μm thickness, calculate the average values of the Poynting vector at A and B for the two frequencies (a) 100 kHz, (b) 100 MHz.

Where does the attenuation occur?

11.6 Show that γ the propagation constant for the TE_{10} mode is

$$\gamma = \left[\left(\frac{\pi}{a} \right)^2 - \left(\frac{\omega}{u_0} \right)^2 \right]^{1/2}$$

and derive expressions for the phase and group velocity of the wave.

$$\left(v_p = u_0 [1 - (\omega/\omega_c)^2]^{-1/2}, \qquad v_g = u_0 [1 - (\omega/\omega_c)^2]^{1/2} \right)$$

11.7 Derive the expressions for the electromagnetic fields of the TM modes in rectangular waveguide.

11.8 Calculate the average value of the Poynting vector along the direction of propagation in a TE_{10} waveguide and the total transmitted power in terms of E_0 the maximum electric field.

$$\left(\frac{E_0^2}{2\omega\mu_0\lambda_g} \sin^2 \left(\frac{\pi x}{a} \right) a_z, \qquad \frac{E_0^2 ab}{4u_0\mu_0} \left[1 - \left(\frac{\lambda_0}{2a} \right)^2 \right]^{1/2} \right)$$

11.9 A coaxial line may be used to support TEM modes. Show that the allowed fields are

$$H_\phi = \frac{1}{\rho} H_0 \exp[j(\omega t - \beta z)]$$

$$E_\rho = Z_0 \frac{1}{\rho} H_0 \exp[j(\omega t - \beta z)]$$

where ρ is the radial coordinate.

11.10 A transverse electric (TE) mode sine wave is propagated along the z axis in a perfect dielectric with a phase constant β_g, whereas the free-space phase constant is β_0. Show that the transverse electric and magnetic fields are given by

$$H_y = \frac{jB_g}{\beta_0^2 - \beta_g^2} \frac{\partial H_z}{\partial y}, \qquad H_x = \frac{-jB_g}{\beta_0^2 - \beta_g^2} \frac{\partial H_z}{\partial x}$$

$$E_y = \frac{j\omega\mu}{\beta_0^2 - \beta_g^2} \frac{\partial H_z}{\partial y}, \qquad E_x = \frac{-j\omega\mu}{\beta_0^2 - \beta_g^2} \frac{\partial H_z}{\partial x}$$

Appendix 1: The Three Main Coordinate Systems

These three systems are only instances of more general *curvilinear* coordinates, which are not used in this text. Discussions of curvilinear coordinates can be found in many of the books listed on p. 280 including that by Hayt (1967).

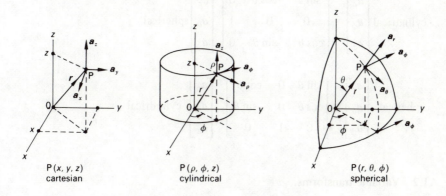

Figure A1.1 *The three main coordinate systems.*

Only in the cartesian system do all three unit datum vectors remain in the same directions; in the others a_ϕ, a_ρ, a_θ, a_r all move with the position.

When changing from one system to another, usually for an easing of the problem, two steps are involved, which can be taken in either order.

(i) Change the components and datum vectors.
(ii) Change the variables.

A1.1 Component transform matrices

$$\text{cartesian} \begin{bmatrix} a_x \\ a_y \\ a_z \end{bmatrix} = \begin{bmatrix} \cos\phi & -\sin\phi & 0 \\ \sin\phi & \cos\phi & 0 \\ 0 & 0 & 1 \end{bmatrix} \begin{bmatrix} a_\rho \\ a_\phi \\ a_z \end{bmatrix} \text{cylindrical}$$

$$\text{cylindrical} \begin{bmatrix} a_\rho \\ a_\phi \\ a_z \end{bmatrix} = \begin{bmatrix} \cos\phi & \sin\phi & 0 \\ -\sin\phi & \cos\phi & 0 \\ 0 & 0 & 1 \end{bmatrix} \begin{bmatrix} a_x \\ a_y \\ a_z \end{bmatrix} \text{cartesian}$$

$$\text{cartesian} \begin{bmatrix} a_x \\ a_y \\ a_z \end{bmatrix} = \begin{bmatrix} (\sin\theta\cos\phi) & (\cos\theta\cos\phi) & -\sin\phi \\ (\sin\theta\sin\phi) & (\cos\theta\sin\phi) & \cos\phi \\ \cos\phi & -\sin\theta & 0 \end{bmatrix} \begin{bmatrix} a_r \\ a_\theta \\ a_\phi \end{bmatrix} \text{spherical}$$

$$\text{spherical} \begin{bmatrix} a_r \\ a_\theta \\ a_\phi \end{bmatrix} = \begin{bmatrix} (\sin\theta\cos\phi) & (\sin\theta\sin\phi) & \cos\theta \\ (\cos\theta\cos\phi) & (\cos\theta\sin\phi) & -\sin\theta \\ -\sin\phi & \cos\phi & 0 \end{bmatrix} \begin{bmatrix} a_x \\ a_y \\ a_z \end{bmatrix} \text{cartesian}$$

$$\text{cylindrical} \begin{bmatrix} a_\rho \\ a_\phi \\ a_z \end{bmatrix} = \begin{bmatrix} \sin\theta & \cos\theta & 0 \\ 0 & 0 & 1 \\ \cos\theta & -\sin\theta & 0 \end{bmatrix} \begin{bmatrix} a_r \\ a_\theta \\ a_\phi \end{bmatrix} \text{spherical}$$

$$\text{spherical} \begin{bmatrix} a_r \\ a_\theta \\ a_\phi \end{bmatrix} = \begin{bmatrix} \sin\theta & 0 & \cos\theta \\ \cos\theta & 0 & -\sin\theta \\ 0 & 1 & 0 \end{bmatrix} \begin{bmatrix} a_\rho \\ a_\phi \\ a_z \end{bmatrix} \text{cylindrical}$$

A1.2 Variable transforms

Table A1.1

	Cartesian	Cylindrical	Spherical
x		$\rho\cos\phi$	$r\sin\theta\cos\phi$
y		$\rho\sin\phi$	$r\sin\theta\sin\phi$
z		z	$r\cos\theta$
	$(x^2+y^2)^{1/2}$	ρ	$r\sin\theta$
	$\tan^{-1}(y/x)$	ϕ	ϕ
	$(x^2+y^2+z^2)^{1/2}$	$(\rho^2+z^2)^{1/2}$	r
	$\tan^{-1}\{(x^2+y^2)^{1/2}/z\}$	$\tan^{-1}(\rho/z)$	θ
d(vol)	$dx\,dy\,dz$	$\rho\,dl\,d\phi\,dz$	$r^2\sin\theta\,dr\,d\theta\,d\phi$

As an example suppose we wish to transform the vector F given in cartesians as $F = F_x a_x + F_y a_y + F_z a_z$ to cylindricals.

A1.2.1   Change components and datum vectors

$$[F_x F_y F_z] \begin{bmatrix} a_x \\ a_y \\ a_z \end{bmatrix} = [F_x F_y F_z] \begin{bmatrix} \cos\phi & -\sin\phi & 0 \\ \sin\phi & \cos\phi & 0 \\ 0 & 0 & 1 \end{bmatrix} \begin{bmatrix} a_\rho \\ a_\phi \\ a_z \end{bmatrix}$$

$$= (F_x \cos\phi + F_y \sin\phi)a_\rho + (-F_x \sin\phi + F_y \cos\phi)a_\phi + F_z a_z$$

$$= F_\rho a_\rho + F_\phi a_\phi + F_z a_z$$

A1.2.2   Change variables x, y, z to ρ, ϕ, z

Suppose a vector field F may be described by

$$F = -\frac{y}{x^2 + y^2} a_x + \frac{x}{x^2 + y^2} a_y + c a_z$$

$$F_\rho = F_x \cos\phi + F_y \sin\phi = \frac{-y}{x^2 + y^2} \cos\phi + \frac{x}{x^2 + y^2} \sin\phi$$

$$= -\frac{\rho \sin\phi}{\rho^2} \cos\phi + \frac{\rho \cos\phi}{\rho^2} \sin\phi = 0$$

$$F_\phi = -F_x \sin\phi + F_y \cos\phi = \frac{y}{x^2 + y^2} \sin\phi + \frac{x}{x^2 + y^2} \cos\phi$$

$$= \frac{\rho \sin\phi}{\rho^2} \sin\phi + \frac{\rho \cos\phi}{\rho^2} \cos\phi = \frac{1}{\rho}$$

$$F_z = c$$

Finally

$$F = 0 a_\rho + \frac{1}{\rho} a_\phi + c a_z$$

Note that the above approach is a general 'sledge-hammer'; often the physical origin of the problem suggests the best coordinate system to use for simplicity, thus avoiding the need for transformation. The above example arises from the magnetic field of a current filament in the z axis plus a uniform field in the z direction.

Appendix 2: Useful Data

This appendix is particularly useful for distribution as a 'hand-out' in examinations.

A2.1   Fundamental physical constants

Permeability in vacuum	μ_0	$4\pi \times 10^{-7}$ H/m
Permittivity in vacuum	ϵ_0	$8.854 \times 10^{-12} \approx 1/36\pi\,10^9$ F/m
Velocity of light	$c(u_0)$	$2.998 \times 10^8 \approx 3 \times 10^8$ m/s
Electronic charge	$e(q_e)$	1.602×10^{-19} C
Electronic rest mass	$m(m_e)$	9.108×10^{-31} kg
Planck's constant	h	6.626×10^{-34} J s
1 ångström unit	Å	1×10^{-10} m
Gravitational constant	g	9.81 m/s^2
Avogadro's number	N_A	6.02×10^{23} mol^{-1}
Boltzmann's constant	k	1.380×10^{-23} J/K

A2.2   Vector identities

Here A and B represent vector field functions and S and T are scalar field functions.

(1) $(A \times B) \cdot C \equiv (B \times C) \cdot A \equiv (C \times A) \cdot B$

(2) $A \times (B \times C) \equiv (A \cdot C)B - (A \cdot B)C$

(3) $\nabla \cdot (A + B) \equiv \nabla \cdot A + \nabla \cdot B$

(4) $\nabla(S + T) \equiv \nabla S + \nabla T$

(5) $\nabla \times (A + B) \equiv \nabla \times A + \nabla \times B$

(6) $\nabla \cdot (SA) \equiv A \cdot \nabla S + S \nabla \cdot A$

(7) $\nabla(ST) \equiv S\nabla T + T\nabla S$

(8) $\nabla \times (SA) \equiv \nabla S \times A + S \nabla \times A$

(9) $\nabla . (A \times B) \equiv B . \nabla \times A - A . \nabla \times B$

(10) $\nabla (A . B) \equiv (A . \nabla)B + (B . \nabla)A + A \times (\nabla \times B) + B \times (\nabla \times A)$

(11) $\nabla \times (A \times B) \equiv A \nabla . B - B \nabla . A + (B . \nabla)A - (A . \nabla)B$

(12) $\nabla . \nabla S \equiv \nabla^2 S$

(13) $\nabla . \nabla \times A \equiv 0$

(14) $\nabla \times \nabla S \equiv 0$

(15) $\nabla \times \nabla \times A \equiv \nabla (\nabla . A) - \nabla^2 A$

A2.3   Vector operations

A2.3.1   Divergence

Cartesian $\nabla . F = \dfrac{\partial F_x}{\partial x} + \dfrac{\partial F_y}{\partial y} + \dfrac{\partial F_z}{\partial z}$

Cylindrical $\nabla . F = \dfrac{1}{\rho} \dfrac{\partial}{\partial \rho} (\rho F_\rho) + \dfrac{1}{\rho} \dfrac{\partial F_\phi}{\partial \phi} + \dfrac{\partial F_z}{\partial z}$

Spherical $\nabla . F = \dfrac{1}{r^2} \dfrac{\partial}{\partial r} (r^2 F_r) + \dfrac{1}{r \sin \theta} \dfrac{\partial}{\partial \theta} (\sin \theta F_\theta) + \dfrac{1}{r \sin \theta} \dfrac{\partial F_\phi}{\partial \phi}$

A2.3.2   Gradient

Cartesian $\nabla S = \dfrac{\partial S}{\partial x} a_x + \dfrac{\partial S}{\partial y} a_y + \dfrac{\partial S}{\partial z} a_z$

Cylindrical $\nabla S = \dfrac{\partial S}{\partial \rho} a_\rho + \dfrac{1}{\rho} \dfrac{\partial S}{\partial \phi} a_\phi + \dfrac{\partial S}{\partial z} a_z$

Spherical $\nabla S = \dfrac{\partial S}{\partial r} a_r + \dfrac{1}{r} \dfrac{\partial S}{\partial \theta} a_\theta + \dfrac{1}{r \sin \theta} \dfrac{\partial S}{\partial \phi} a_\phi$

A2.3.3   Curl

Cartesian $\nabla \times F = \left(\dfrac{\partial F_z}{\partial y} - \dfrac{\partial F_y}{\partial z} \right) a_x + \left(\dfrac{\partial F_x}{\partial z} - \dfrac{\partial F_z}{\partial x} \right) a_y + \left(\dfrac{\partial F_y}{\partial x} - \dfrac{\partial F_x}{\partial y} \right) a_z$

Cylindrical $\nabla \times F = \left(\dfrac{1}{\rho} \dfrac{\partial F_z}{\partial \phi} - \dfrac{\partial F_\phi}{\partial z} \right) a_\rho + \left(\dfrac{\partial F_\rho}{\partial z} - \dfrac{\partial F_z}{\partial \rho} \right) a_\phi + \dfrac{1}{\rho} \left(\dfrac{\partial (\rho F_\phi)}{\partial \rho} - \dfrac{\partial F_\rho}{\partial \phi} \right) a_z$

Spherical $\nabla \times F = \dfrac{1}{r \sin \theta} \left(\dfrac{\partial (F_\phi \sin \theta)}{\partial \theta} - \dfrac{\partial F_\theta}{\partial \phi} \right) a_r + \dfrac{1}{r} \left(\dfrac{1}{\sin \theta} \dfrac{\partial F_r}{\partial \phi} - \dfrac{\partial (r F_\phi)}{\partial r} \right) a_\theta$

$\qquad\qquad + \dfrac{1}{r} \left(\dfrac{\partial (r F_\theta)}{\partial r} - \dfrac{\partial F_r}{\partial \theta} \right) a_\phi$

A2.3.4　Laplacian

Cartesian $\nabla^2 S = \dfrac{\partial^2 S}{\partial x^2} + \dfrac{\partial^2 S}{\partial y^2} + \dfrac{\partial^2 S}{\partial z^2}$

Cylindrical $\nabla^2 S = \dfrac{1}{\rho}\dfrac{\partial}{\partial \rho}\left(\rho\dfrac{\partial S}{\partial \rho}\right) + \dfrac{1}{\rho^2}\dfrac{\partial^2 S}{\partial \phi^2} + \dfrac{\partial^2 S}{\partial z^2}$

Spherical $\nabla^2 S = \dfrac{1}{r^2}\dfrac{\partial}{\partial r}\left(r^2\dfrac{\partial S}{\partial r}\right) + \dfrac{1}{r^2 \sin\theta}\dfrac{\partial}{\partial \theta}\left(\sin\theta\dfrac{\partial S}{\partial \theta}\right) + \dfrac{1}{r^2 \sin^2\theta}\dfrac{\partial^2 S}{\partial \phi^2}$

Appendix 3: Solid Angle

A knowledge of solid angle is sometimes needed for problems involving the magnetic scalar potential U.

(i) The complete surface of a sphere, of any radius r, subtends 4π units of solid angle at its centre. A unit is called a *steradian*. This is shown in figure A3.1a. It follows that the surface also subtends 4π steradians at any other point P within the sphere; see figure A3.1b. For a part of the surface S, shown in diagram A.3.1c, the solid angle Ω subtended at the centre is S/r^2 steradians.

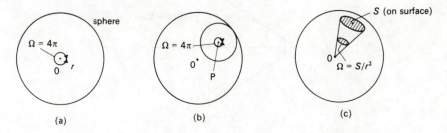

Figure A3.1 *Definition of solid angle Ω with reference to a sphere.*

(ii) The vector area $d\mathbf{S}$ which subtends a point P has a solid angle given by $d\Omega = d\mathbf{S}.\,\mathbf{a}_r/r^2$. It is the area subtended on a sphere of unit radius centred at P, see figure A3.2a.

(iii) The solid angle subtended at any point P is given by the surface integral

$$\Omega = \int_S d\mathbf{S}.\,\mathbf{a}_r/r^2 \text{ steradian}$$

The surface is not necessarily closed as shown in figure A3.2b.

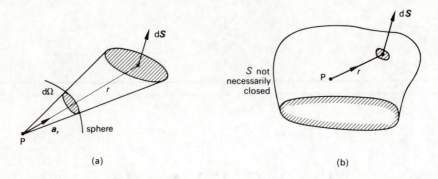

Figure A3.2   *General definition of solid angle Ω.*

A3.1   Examples of solid angles

Example 1   cone of semi-angle α   It is a straightforward matter to show that the surface area of the enclosing sphere cut off by the base of the cone, that is the *cap area*, is $2\pi r^2 (1 - \cos \alpha)$. This is shown in figure A3.3.

Hence solid angle $\Omega = S/r^2 = 2\pi(1 - \cos \alpha)$ steradians.

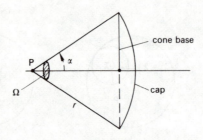

Figure A3.3   *Solid angle subtended by a cone.*

Example 2   plane sheets wholly, semi- and quarter infinite   These are illustrated in figure A3.4a, b and c, shown in the xy plane. The point P considered is on the z axis, a little behind the origin. The portions of the sphere through P cut off (and removed) by the planes show that the corresponding solid angles they subtend at P are 2π, π and $\pi/2$ steradians respectively.

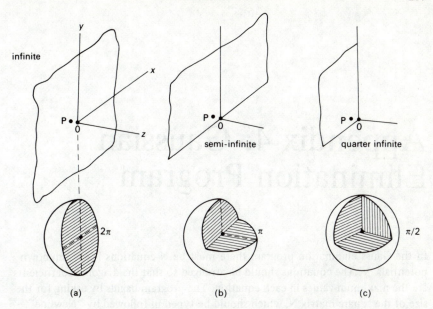

Figure A3.4   *Solid angle subtended by planes.*

Appendix 4: Gaussian Elimination Program

In the Gauss elimination program there must be N equations for N unknown potentials, V. The equations should be arranged so that the diagonal coefficients are the maximum values in each equation. The program begins by asking for the size of the square matrix N, which should be typed in followed by "Newline". The coefficients S(N,N) are then asked for, followed by a request for the values of the column matrix R(N). The solutions are then printed out in reverse order $Vn, Vn-1, \ldots, V2, V1$.

As a simple example, from chapter 8 we have the set of equations

$$-4V_1 + 4V_2 + 0V_3 = 5$$

$$+V_1 - 4V_2 + 2V_3 = 5$$

$$0V_1 + 2V_2 - 4V_3 = 3$$

These would be input as N = 3

S(N,N)			R(N)
−4	4	0	5
+1	−4	2	5
0	2	−4	3

The output is $-3.4375, -4.375, -5.625$.

Once the data have been loaded into arrays S(M,P) and R(P), the unknown potentials are printed out automatically. However, if it is required to change data or if a mistake has been made in loading data, the programme can be re-run by using GOTO 100 without destroying the main body of data. Corrections can easily be made by typing

LET S(M,P) = ?

or

LET R(P) = ?

260

If a listing is required of these array contents, then use GOTO 1500.

Comments

"Asks for size of square matrix"

```
30    REM "GAUSS ELIMINATION"
40    PRINT "N?";
45    INPUT N
50    PRINT "N =";N
52    PRINT "S(N,N) =";
53    DIM S(N,N)
54    FOR M = 1 TO N
55    FOR P = 1 TO N
61    INPUT S(M,P)
62    PRINT S(M,P);",",";
66    NEXT P
67    NEXT M
70    DIM R(N)
72    PRINT "R(N) = ";
74    FOR Q = 1 TO N
85    INPUT R(Q)
86    PRINT R(Q);",",";
89    NEXT Q

100   DIM Y(N,N)
110   DIM I(N)
115   DIM V(N)
120   FOR M = 1 TO N
140   FOR P = 1 TO N
141   LET X = S(M,P)
143   LET Y(M,P) = X
150   LET Z = R(M)
155   LET I(M) = Z
240   NEXT P
260   NEXT M
900   PRINT "U(B) = ";
1000  FOR A = 2 TO N
1020  LET B = A−1
1040  FOR C = A TO N
1060  LET PIV = Y(C,B)/Y(BB)
1080  IF PIV = 0 THEN GOTO 1170
1100  FOR D = B TO N
1120  LET Y(C,D) = Y(C,D)/PIV−Y(B,D)
1140  NEXT D
```

"Input coefficients in serial form (these are saved in array S(N,N) and are not destroyed). The program employs a 'live' array Y(N,N) which cannot be used for inputting data"

"Input column matrix in serial form. Again, data is not destroyed but replaced by a 'live' array I(M) in the program"

"Operating 'live' arrays"

"This is the main part of the program"

```
1160    LET I(C) = I(C)/PIV−I(B)
1170    NEXT C
1175    NEXT A
1200    FOR A = 1 TO N
1220    LET B = N + 1−A
1240    LET C = B + 1
1260    LET U = 0
1280    IF A = 1 THEN GOTO 1360
1300    FOR D = C TO N
1320    LET U = U + Y(B,D) ∗ V(D)
1340    NEXT D
1360    LET V(B) = (I(B)−U)/Y(B,B)
1370    PRINT V(B);".";
1380    NEXT A
1400    STOP

1500    FOR M = 1 TO N
1520    FOR P = 1 TO N
1540    PRINT S(M,P)
1560    NEXT P
1570    PRINT
1580    NEXT M
1585    FOR P = 1 TO N
1590    PRINT R(P)
1595    NEXT P
1600    STOP
```

"By using GOTO 1500 this routine provides a listing of the data in the square array S(M,P) and the column array R(P)"

Solutions to Selected Problems

1.1   (a)

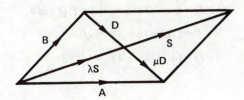

Figure S1.1(a).

Let λ and μ be two real scalar quantities < 1. We have to prove that $\lambda = \mu = \frac{1}{2}$. From the vector diagram in figure S1.1(a)

$$A = \lambda S + \mu D = \lambda(A + B) + \mu(A - B)$$

$$= A(\lambda + \mu) + B(\lambda - \mu)$$

Therefore

$$A(1 - \lambda - \mu) = B(\lambda + \mu)$$

Since A and B are two vectors *not* in the same direction, *both* coefficients must be zero. Therefore

$$1 - \lambda - \mu = 0$$

$$\lambda + \mu = 0$$

The solution of these two equations yields

$$\lambda = \tfrac{1}{2}, \ \mu = \tfrac{1}{2} \qquad\qquad \text{Q.E.D.}$$

263

(b)
$$P = 1i + 3j - 7k = OP$$
$$Q = 5i - 2j + 4k = OQ$$
$$PQ = OQ - OP = 4i - 5j + 11k$$

A unit vector in direction PQ is then

$$a_{PQ} = (4i - 5j + 11k)/(4^2 + 5^2 + 11^2)^{\frac{1}{2}}$$

Therefore

$$a_{PQ} = (4i - 5j + 11k)/9\sqrt{2}$$

1.2   (a) (i) In this case we have

$$P = 4i + 6j + 2k$$
$$Q = 2i - 12j + 2k$$

From the vector diagram note that $\theta > 90°$. Also

$$P.Q = 8 - 72 + 4 = -60 = P.Q \cos \theta$$
$$P = (16 + 36 + 4)^{\frac{1}{2}} = \sqrt{56}$$
$$Q = (4 + 144 + 4)^{\frac{1}{2}} = \sqrt{152}$$

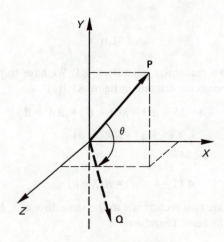

Therefore

$$\cos \theta = -\frac{60}{\sqrt{152} \times \sqrt{56}} = -0.6503$$

$$\underline{\theta = 130.6°}$$

This angle may be checked by $P \times Q = PQ \sin \theta \, a_n$

$$P \times Q = \begin{vmatrix} i & j & k \\ 4 & 6 & 2 \\ 2 & -12 & 2 \end{vmatrix} = 4 \, (9i - 1j - 15k)$$

Therefore

$$PQ \sin \theta = 4 \, (9^2 + 1^2 + 15^2)^{\frac{1}{2}} = 4 \sqrt{307}$$

Hence

$$\sin \theta = \frac{4\sqrt{307}}{PQ} = 0.7596$$

and hence

$$\underline{\theta = 180 - 49.4 = 130.6°}$$

(a) (ii) The unit vector a_n is

$$a_n = \pm \, (9i - 1j - 15k)/\sqrt{307}$$

(b) To prove $(A \times B) . (C \times D) = (A.C) \, (B.D) - (A.D) \, (B.C)$.

$$(A \times B) . (C \times D) = (A \times B) . E \text{ (say)}$$
$$= A . (B \times E)$$
$$= A . [B \times (C \times D)]$$
$$= A . [C(B.D) - D(B.C)]$$

Since $B \times (C \times D)$ is a vector triple product, see section 1.2.6, therefore

$$(A \times B) . (C \times D) = (A.C) \, (B.D) - (A.D) \, (B.C) \qquad \text{Q.E.D.}$$

1.5 $\qquad\qquad r = xi + yj + zk; \; |r| = [(x^2 + y^2 + z^2)]^{\frac{1}{2}}$

(i) $\operatorname{div} r = \nabla . r = \dfrac{\partial r_x}{\partial x} + \dfrac{\partial r_y}{\partial y} + \dfrac{\partial r_z}{\partial z} = 1 + 1 + 1 = 3$

(ii) $\operatorname{curl} r = \nabla \times r = \begin{vmatrix} i & j & k \\ \partial/\partial x & \partial/\partial y & \partial/\partial z \\ x & y & z \end{vmatrix}$

$$= i(0) + j(0) + k(0) = 0$$

(iii) $-\mathrm{grad}\left(\dfrac{1}{r}\right) = - \nabla [x^2 + y^2 + z^2]^{-\frac{1}{2}}$

$$= -\left[\frac{\partial}{\partial x}i + \frac{\partial}{\partial y}j + \frac{\partial}{\partial z}k\right]F(x, y\, z)$$

$$= -\left\{\left(-\frac{1}{2}\right)2x\ [\ ]^{-3/2}i + \left(-\frac{1}{2}\right)2y\ [\ ]^{-3/2}j + \left(-\frac{1}{2}\right)2z\ [\ ]^{-3/2}k\right\}$$

$$= [\ ]^{-3/2}\,[xi + yj + zk] = r/|r|^3$$

(iv) $\nabla^2 |r|^2 = \left[\dfrac{\partial^2}{\partial x^2} + \dfrac{\partial^2}{\partial y^2} + \dfrac{\partial^2}{\partial z^2}\right](x^2 + y^2 + z^2)$

$$= 2 + 2 + 2 = 6$$

(v) $\mathrm{grad}\,(A.r) = \nabla(A.r)$

$$= \nabla(A_x i + A_y j + A_z k)\,.\,(xi + yj + zk)$$

$$= \nabla[xA_x + yA_y + zA_z]$$

$$= \left(\frac{\partial}{\partial x}i + \frac{\partial}{\partial y}j + \frac{\partial}{\partial z}k\right)[xA_x + yA_y + zA_z]$$

$$= A_x i + A_y j + A_z k = \mathbf{A}$$

which is a *constant* vector.

(vi) Curl $(A \times r)$

$$A \times r = \begin{vmatrix} i & j & k \\ A_x & A_y & A_z \\ x & y & z \end{vmatrix}$$

$$= i\,(zA_y - yA_z) + j\,(xA_z - zA_x) + k\,(yA_x - xA_y)$$

$$\mathrm{Curl}\,(A \times r) = \begin{vmatrix} i & j & k \\ \partial/\partial x & \partial/\partial y & \partial/\partial z \\ (zA_y - yA_z) & (xA_z - zA_x) & (yA_x - xA_y) \end{vmatrix}$$

$$= i\,[A_x + A_x] + j\,[A_y + A_y] + k\,[A_z + A_z]$$

$$= 2A$$

1.6 $\quad F = xi + 2yj + 3zk;\ S = (x^2 - y^2 + z^2)$

$$\mathrm{div}\,SF = \nabla.(SF) = \frac{\partial}{\partial x}(SF_x) + \frac{\partial}{\partial y}(SF_y) + \frac{\partial}{\partial z}(SF_z)$$

$$= \left[S \frac{\partial F_x}{\partial x} + F_x \frac{\partial S}{\partial x} \right] + \left[S \frac{\partial F_y}{\partial y} + F_y \frac{\partial S}{\partial y} \right] + \left[S \frac{\partial F_z}{\partial z} + F_z \frac{\partial S}{\partial z} \right]$$

$$= S \left(\frac{\partial F_x}{\partial x} + \frac{\partial F_y}{\partial y} + \frac{\partial F_z}{\partial z} \right) + F_x \frac{\partial S}{\partial x} + F_y \frac{\partial S}{\partial y} + F_z \frac{\partial S}{\partial z}$$

$$= S \operatorname{div} \boldsymbol{F} + \boldsymbol{F} . \operatorname{grad} S = S \nabla . \boldsymbol{F} + \boldsymbol{F} . \nabla S$$

$$= (x^2 - y^2 + z^2)(1 + 2 + 3) + (xi + 2yj + 3zk) . (2xi - 2yj + 2zk)$$

$$= 6(x^2 - y^2 + z^2) + (2x^2 - 4y^2 + 6z^2)$$

$$= 8x^2 - 10y^2 + 12z^2$$

At $(2, 2\sqrt{2}, 2)$

$$\operatorname{div}(SF) = 32 - 80 + 48 = 0$$

At $(1, 1, 1)$

$$\operatorname{div}(SF) = 8 - 10 + 12 = +10$$

Hence, the vector field *SF* is *not* everywhere of zero divergence and therefore it cannot be solenoidal.

2.1

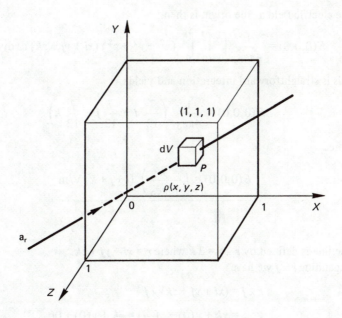

Figure S2.1.

The charge contained in an element of volume dV shown at P in figure S2.1 is

$$\rho_V \, dV = - (x^2 + y^2 + z^2)^{5/2} \, dxdydz$$

The electric field is

$$dE = \frac{\rho_V \, dV \, (-a_r)}{4\pi\epsilon_0 r^2}$$

(this is negative because we have to compute E at the origin) where

$$r = xi + yj + zk$$

and

$$a_r = \frac{xi + yj + zk}{(x^2 + y^2 + z^2)^{1/2}}$$

Substituting for a_r and r in dE gives

$$dE = + \frac{(x^2 + y^2 + z^2)^{5/2} \, dxdydz}{4\pi\epsilon_0 (x^2 + y^2 + z^2)} \cdot \frac{xi + yj + zk}{(x^2 + y^2 + z^2)^{1/2}}$$

$$= \left(\frac{1}{4\pi\epsilon_0}\right) \, [(x^2 + y^2 + z^2) \, (xi + yj + zk) \, dxdydz]$$

The electric field at the origin is then

$$E(0,0,0) = \frac{1}{4\pi\epsilon_0} \int_0^1 \int_0^1 \int_0^1 (x^2 + y^2 + z^2) \, (xi + yj + zk) \, dxdydz$$

This is straightforward integration and yields

$$E(0,0,0) = \frac{1}{4\pi\epsilon_0} \left(\frac{7}{12} i + \frac{7}{12} j + \frac{7}{12} k\right)$$

Hence

$$E(0,0,0) = \left(\frac{7}{48\pi\epsilon_0}\right) \, (i + j + k) \, V/m$$

2.2   The line is defined by $r \times j = 3k$ where $r = xi + yj + zk$.
       Expanding $r \times j$ we have

$$r \times j = (xi + yj + zk) \times j$$

$$= xk + y(0) + z(-i) = 3k + y(0) + 0i$$

Equating gives

$$x = 3, \ y = y, \ z = 0$$

Alternatively

$$r \times j = xk - zi = 3k$$

Therefore

$$xk - 3k = zi$$

$$(x - 3)k = zi$$

Since k and i are not in the same direction, then

$$x - 3 = 0, \ z = 0$$

Therefore

$$r = 3i + yj$$

This may be sketched as shown in figure S2.2(a).
For an infinite line charge $+\rho_L$

$$E = \left(\frac{\rho_L}{2\pi\epsilon_0}\right) \ \left(\frac{1}{a}\right) a_\rho$$

where $a = 3$. The field at $(0, 5, 0)$ is then

$$\underline{E_1 = -12 \, i \, \text{V/m}}$$

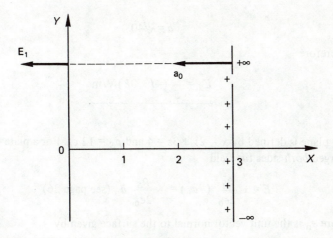

Figure S2.2(a).

The diagram for the point $P_2 = (1, 2, 6)$ is shown in figure S2.2(b).

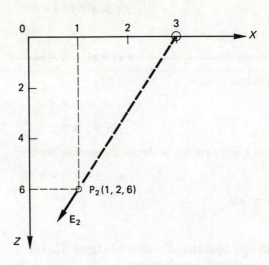

Figure S2.2(b).

As in the previous case the y dimension has no influence since the line is infinite in y. The unit vector a_ρ is now

$$a_\rho = \frac{-2i + 6k}{\sqrt{40}} = \frac{-i + 3k}{\sqrt{10}}$$

and

$$a = \sqrt{40}$$

Therefore

$$E_2 = \frac{9}{5} \, (-i + 3k) \, \text{V/m}$$

2.3   The plane is defined by $x - 2y + 3z = 4$ and $\rho_S = 12 \, \epsilon_0$. For a plate uniformly charged *both* sides the field

$$E = + \frac{\rho_S}{2\epsilon_0} \, (-a_n) = - \frac{\rho_S}{2\epsilon_0} \, a_n \text{ (see page 26)}$$

where a_n is the unit vector normal to the surface given by

$$a_n = \frac{\text{grad } S}{|\text{grad } S|}$$

$$\text{grad } S = \left(\frac{\partial}{\partial x} \, i + \frac{\partial}{\partial y} \, j + \frac{\partial}{\partial z} \, k \right) \, (x - 2y + 3z)$$

$$= 1i - 2j + 3k$$

Therefore

$$a_n = (i - 2j + 3k)/(1 + 4 + 9)^{\frac{1}{2}}$$

$$= (i - 2j + 3k)/\sqrt{14}$$

hence

$$E = \frac{12\epsilon_0}{2\epsilon_0} \, (i - 2j + 3k)/\sqrt{14}$$

and therefore

$$\underline{E = -6 \, [i - 2j + 3k] /\sqrt{14} \text{ V/m}}$$

4.3 (a) $J = J_0 \exp(-|\rho|/a) \, a_z \text{ A/m}^2$.

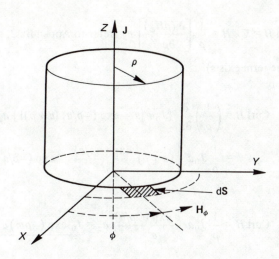

This has cylindrical symmetry and the total current is

$$I = \int_S J \cdot dS \quad \text{where} \quad dS = 2\pi\rho d\rho \, a_z$$

Therefore

$$I = \int_S J_0 \exp(-\rho/a)\,\boldsymbol{a}_z \cdot 2\pi\rho\,\mathrm{d}\rho\,\boldsymbol{a}_z = \int_S J_0\,2\pi\rho \exp(-\rho/a)\,\mathrm{d}\rho$$

$$= -2\pi J_0 a \left[\rho \exp(-\rho/a) - \int_S \exp(-\rho/a)\,\mathrm{d}\rho\right]$$

after integration by parts.

Integrating over a surface between $\rho = 0$ and $\rho = \rho$ gives

$$I = -2\pi J_0\, a\,[\rho \exp(-\rho/a) + a \exp(-\rho/a) - a]$$

$$= 2\pi J_0\, a\,[a - \exp(-\rho/a)\,(a + \rho)]$$

(b) There is cylindrical symmetry so that the H_ϕ component only exists. Using Ampère's circuital law gives

$$I = \oint H.\mathrm{d}l = H_\phi\,2\pi\rho = 2\pi J_0 a\,[a - \exp(-\rho/a)\,(a+\rho)]$$

Hence

$$H = H_\phi\,\boldsymbol{a}_\phi = \frac{J_0 a}{\rho}\,[a - \exp(-\rho/a)\,(a+\rho)]\,\boldsymbol{a}_\phi$$

(c) Curl $H = \nabla \times H = \dfrac{1}{\rho}\left[\dfrac{\partial\,(\rho H_\phi)}{\partial\rho}\right]\boldsymbol{a}_z$ (refer to Appendix 2, section A2.3.3,

only one term exists).
Therefore

$$\mathrm{Curl}\,H = \left(\frac{1}{\rho}\right)\frac{\partial}{\partial\rho}\,\{J_0 a\,[a - \exp(-\rho/a)\,(a+\rho)]\}\,\boldsymbol{a}_z$$

$$= \frac{1}{\rho}\,J_0 a\left\{(a+\rho)\,\frac{\exp(-\rho/a)}{a} - \exp(-\rho/a)\right\}\boldsymbol{a}_z$$

hence

$$\mathrm{Curl}\,H = \frac{1}{\rho}\,J_0 a\,\frac{\rho \exp(-\rho/a)}{a}\,\boldsymbol{a}_z = \underline{J_0 \exp(-\rho/a)\,\boldsymbol{a}_z} \qquad\qquad \text{Q.E.D.}$$

5.1 (a) $B = (5i + 2j - 4k)/10$

   $I = 50$ A

   The force is given by

$$F = Il \times B$$

where $l_1 = 1i,\ l_2 = 2j,\ l_3 = -1i,\ l_4 = -2j.$

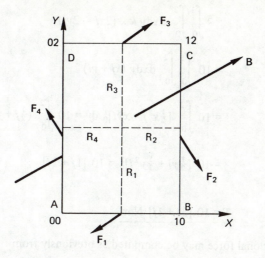

Figure S5.1(a).

Therefore

$$F_1 = 5i \times (5i + 2j - 4k) = 5(0 + 2k + 4j)$$

$$F_2 = 5(2j) \times (5i + 2j - 4k) = 10(-5k + 0 - 4i)$$

$$F_3 = -F_1 \qquad\qquad = 5(0 - 2k - 4j)$$

$$F_4 = -F_2 \qquad\qquad = 10(5k + 0 + 4i)$$

The total force is $F = F_1 + F_2 + F_3 + F_4 = 0$, that is no translational force.
The total torque is

$$T = IS \times B$$

where S is the area $= 1i \times 2j = 2k$.
Therefore

$$T = 5(2k) \times (5i + 2j - 4k) = 10[5j - 2i + 0]$$

hence

$$\underline{T = 10(-2i + 5j)\,\text{N m}}$$

(b) $B = (2xi - 2yj)/10$, $dS = dxdyk$
which is an element of area $dxdy$ in figure S5.1(a). The total torque is then

$$T = I \int_S dS \times B$$

$$= 5 \int_0^1 \int_0^2 dxdy \; k \times (2xi - 2yj)$$

$$= 10 \int_0^1 \int_0^2 dxdy \; (xj + yi)$$

$$= 10 \int_0^2 [\tfrac{1}{2}x^2 j + xyi]_0^1 \; dy = 10 \int_0^2 (\tfrac{1}{2}j + yi) \, dy$$

$$= 10 \; [\tfrac{1}{2}yj + \tfrac{1}{2}y^2 i]_0^2 = 10 \; [1j + 2i]$$

Therefore

$$\underline{T = 10 \; [2i + 1j] \; \text{N m}}$$

The translational force may be computed as previously from

$$dF = Idl \times B \quad \text{and} \quad B = (2xi - 2yj)/10$$

$$F_1 = \frac{50}{10} \int_0^1 dxi \times [2xi - 0j] = 0$$

$$F_2 = 5 \int_0^2 dy \, j \times [2i - 2yj] = 5 \int_0^2 -2 \, k \, dy = -20 \, k$$

$$F_3 = 5 \int_0^1 -dx \, i \times [2xi - 2x2j] = 5 \int_0^1 4k dx = 20 \, k$$

$$F_4 = 5 \int_0^1 -dy \, j \times [-2y \, j] = 0$$

Hence, $F = F_1 + F_2 + F_3 + F_4 = 0$ and there is *NO translational force*. Check on T. The total torque is

$$T = R_2 \times F_2 + R_3 \times F_3$$

where $R_2 = \frac{1}{2}i$ and $R_3 = 1j$ since F_1 and F_4 are zero.

Therefore

$$T = \frac{1}{2}i \times (-20 \, k) + 1j \times (20k)$$

$$= \underline{10j + 20i \; \text{N m which agrees.}}$$

(c) $B = (2yi + 2j + 2xk)/10$.

$$T = I \int_S dS \times B = 5 \int_0^1 \int_0^2 dx \, dy \, k \times (2yi + 2j + 2xk)$$

$$= 10 \int_0^1 \int_0^2 (yj - 1i + 0) \, dx dy$$

$$= 10 \int_0^2 [yxj - xi]_0^1 \, dy = 10 \int_0^2 (yj - i) \, dy$$

$$= 10 \left[\tfrac{1}{2} y^2 j - iy \right]_0^2 = 10 \left[-2i + 2j \right]$$

Therefore

$$T = 20 \left[-i + j \right] \text{ N m}$$

For $y = 0, F_1 = 10 \int_0^1 dx \, i \times [j + xk] = 10 \left[k - \dfrac{1}{2} j \right]$

$x = 1, F_2 = 10 \int_0^2 dy \, j \times [yi + j + k] = 10 \left[-2k + 2i \right]$

$y = 2, F_3 = 10 \int_0^1 -dx \, i \times [2i + j + xk] = 10 \left[-k + \dfrac{1}{2} j \right]$

$y = 0, F_4 = 10 \int_0^2 -dy \, j \times [yi + j + 0k] = 10 \left[2k \right] = 20 \, k$

Total translational force is

$$F = F_1 + F_2 + F_3 + F_4 = 20i \text{ N which is NOT ZERO.}$$

Check, using reaction arms, gives $20 [-i + j]$ N m as before.

6.1 For this problem we can use either Faraday's law

$$V_{\text{emf}} = \oint E \cdot dl^* = - \frac{d\Phi}{dt}$$

or

$$V_{\text{emf}} = - \oint_S \frac{\partial B}{\partial t} \cdot dS + \oint_l (u \times B) \cdot dl^*$$

where the asterisk dl^* refers to those parts which are moving. Faraday's law is simpler to apply, but we shall try both initially.

(a) $B = B_0 k$. Consider the two approaches.

(i) *Faraday*
The flux is

$$\Phi = B_0 Mx = B_0 Mut, \text{ Wb}$$

The e.m.f. is

$$V_{emf} = -B_0Mu = -20 \text{ V}$$

(ii) *Alternative method*

$$V_{emf} = -\int_S \frac{\partial B}{\partial t} \cdot dS + \int_l (u \times B) \cdot dl*$$

$$= 0 + \int_l (ui \times B_0k) \cdot dyj$$

$$= \int_l (20i \times k) \cdot dyj = 20 \int_l -j \cdot dyj$$

$$= -20\, M = -20 \text{ V} \quad \text{(meter reads } downscale\text{)}$$

(b) $B = B_0 \exp(-at)k$ where $B_0 = 1, a = 15$.

$$V_{emf} = -\frac{d\Phi}{dt} = -\frac{d}{dt}\,[B_0 \exp(-at).Mut]$$

Therefore

$$V_{emf} = -B_0Mu\,[-at \exp(-at) + \exp(-at)]$$

$$= -B_0Mu \exp(-at)\,[1 - at]$$

$$= -20 \exp(-at)\,[1 - at]$$

$$= -20 \exp\left(-a\,\frac{L}{u}\right)\left[1 - a\,\frac{L}{u}\right]$$

since $ut = L$. Therefore

$$V_{emf} = -20 \exp\left(-15 \times \frac{2}{20}\right)\left[1 - \frac{15 \times 2}{20}\right]$$

$$= +2.23 \text{ V (upscale)}$$

(Note: at $t = 0$
$V_{emf} = -20 \text{ V}$)

(c) $B = B_0 \exp(\beta x)k$ where $\beta = 0.1, B_0 = 1$.

$$V_{emf} = -\frac{d\Phi}{dt} \quad \text{where now we put}$$

$$\Phi = \int_S B \cdot dS = \int B_0 \exp(\beta x)k \cdot Mdxk$$

$$= MB_0 \int_0^L \exp{(\beta x)}\, dx = MB_0\; \frac{1}{\beta}\; [e^{\beta L} - 1]$$

$$= 10(e^{0.1ut} - 1) = 10(e^{2t} - 1)$$

Therefore

$$\frac{d\Phi}{dt} = 20\; e^{2t}$$

hence

$$V_{\text{emf}} = -\; 20\; e^{2t} \text{ and for } t = \frac{L}{u} = \frac{2}{20} = \frac{1}{10}$$

$$\underline{V_{\text{emf}} = -20\; e^{0.2} = -24.4 \text{ V } downscale}$$

(Note: at $t = 0$,

$$\underline{V_{\text{emf}} = -\; 20 \text{ V})}$$

(d) $B = B_0 \cos{(\omega t - \beta x)}k$, $\omega = 20\; \pi$, $\beta = 0.1$, $B_0 = 1$.

Therefore

$$\Phi = \int_S B . dS = \int_S B_0 \cos{(\omega t - x)}k . M dx k$$

$$= M B_0 \int_{x=0}^{x=L} \cos{(\omega t - \beta x)}\, dx$$

Let $\theta = (\omega t - \beta x)$

hence

$$\frac{d\theta}{dx} = -\beta \quad \text{and} \quad dx = -\; \frac{d\theta}{\beta}$$

Therefore

$$\Phi = -\; \frac{1}{\beta} \int_{\theta_1}^{\theta_2} \cos{\theta}\; d\theta = -\; \frac{1}{\beta} [\sin{\theta}]_{\theta_1\,(x=0)}^{\theta_2\,(x=L)}$$

and hence

$$\Phi = -\; \frac{MB_0}{\beta}\; [\sin{(\omega t - \beta L)} - \sin{\omega t}].$$

In this we note that $L = f(t) = ut$, therefore

$$V_{\text{emf}} = -\; \frac{d\Phi}{dt} = \frac{MB_0}{\beta} \left[\frac{d}{dt} \sin{(\omega t - \beta \omega t)} - \frac{d}{dt} (\sin{\omega t}) \right]$$

$$= \frac{MB_0}{\beta}\; [(\omega - \beta u) \cos{(\omega t - \beta L)} - \omega \cos{\omega t}]$$

and hence

$$V_{emf} = -32.1 \text{ V downscale}$$

Note: at $t = 0$

$$V_{emf} = \frac{MB_0}{\beta} [\omega - \beta u - \omega] = -MB_0 u$$

$$= -20 \text{ V})$$

8.1 *Field sketches:* see figure S8.1 opposite.

(a)

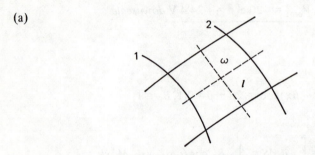

The capacitance between (1) and (2) in the above figure is

$$C_a = \frac{\omega d \epsilon}{l}$$

where d is the depth and ω and l are as shown.

For a unit depth $d = 1$ m, then $\omega = 1$ and

$$C_d = \epsilon$$

The total capacitance is, see the above figure

$$C = C_a \times \frac{\text{No. in parallel}}{\text{No. in series}}$$

$$= \frac{6}{36 \, \pi \times 10^9} \times \frac{4.6 \times 4}{3} = 325.68 \text{ pF/m}$$

or $C = 0.325 \ \mu\text{F/km}$

(b) The greatest stress is along the Y axis. In figure 8.1(b) this is found to be at the *core*.

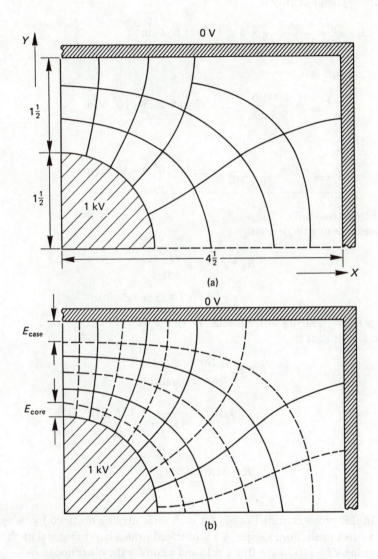

Figure S8.1 *Field sketches: (a) initial sketch, (b) final sketch.*

$$E_{core} = \frac{1000/6}{5.5 \text{ mm}} = 304 \text{ V/cm} = 30.4 \text{ kV/m}$$

The surface charge density is

$$\rho_{s_{core}} = D = \epsilon E = \frac{6}{36\pi \times 10^9} \times 3 \times 10^4 = 1.6 \text{ } \mu C/m^2$$

The stress at the case is

$$E_{case} = \frac{1000/6}{7.5 \text{ mm}} = 223 \text{ V/cm} = 2.23 \times 10^4 \text{ V/m}$$

and

$$\rho_{s_{case}} = D = \frac{6}{36 \times 10^9} \times 2.23 \times 10^4 = 1.18 \text{ } \mu C/m^2$$

(c)*Insulation resistance*
The resistance of a single cell is

$$R_a = \rho \frac{l}{a} = \rho \frac{l}{\omega d}$$

$$= \rho/m \text{ for } l = \omega \text{ and } d = 1 \text{ m}$$

where ρ is the resistivity of the insulation ($\rho = 2 \times 10^8 \text{ } \Omega \text{ m}$).
The total resistance is

$$R = R_a \frac{\text{No. in series}}{\text{No. in parallel}}$$

$$= 2 \times 10^8 \times \frac{3}{4 \times 4.6} = 32.6 \times 10^6 \text{ } \Omega/m$$

or

$$R = 32.6 \text{ k}\Omega/\text{km}$$

8.3   In this problem, initial values for each node may be obtained by using
a larger mesh. For example, V5 is obtained from a larger diagonal mesh
giving V5 = (100 + 0 + 0) / 4 = 25 and so on for the other nodes. A
simple iterative program is given below which for 21 iterations gives
the output shown. Note that in order to estimate nodes 1, 2 and 3 the
nodes immediately above these have been assigned values corresponding
to nodes 4, 5 and 6. Similarly, to estimate node 11, the node to the
right of 11 has been assigned the value of V10. This is possible because
of the symmetry of the problem. The Gauss elimination solution using

the program in Appendix 4 is also shown along with the input data format. The agreement between the two methods is excellent. However, the iterative technique is very much simpler in terms of programming the computer.

```
20 REM "IT 8.3"
30 LET V1 = 6
40 LET V2 = 11
50 LET V3 = 75
60 LET V4 = 9
70 LET V5 = 25
80 LET V6 = 64
90 LET V7 = 6
100 LET V8 = 16.9
110 LET V9 = 31.25
120 LET V10 = 45
130 LET V11 = 50
140 FOR C = 1 TO 20

200 LET V1 = (V2 + 2 * V4)/4
210 LET V2 = (V3 + V1 + 2 * V5)/4
220 LET V3 = (100 + V2 + 2 * V6)/4
230 LET V4 = (V5 + V1 + V7)/4
240 LET V5 = (V6 + V2 + V4 + V8)/4
250 LET V6 = (100 + V3 + V5 + V9)/4
260 LET V7 = (V8 + V4)/4
270 LET V8 = (V9 + V5 + V7)/4
280 LET V9 = (V10 + V6 + V8)/4
290 LET V10 = (V11 + 100 + V9)/4
300 LET V11 = (100 + 2 * V10)/4
340 PRINT "C=" ; C,, V1,, V2,,
V3,, V4,, V5,, V6,, V7,, V8,, V9,,
V10,, V11,, "***"
360 NEXT C
```

Output after 21 iterations	*Gaussian elimination*
14.844779	14.84445
33.60086	33.600489
60.765063	60.764852
12.888881	12.888654
29.396584	29.396328
54.725608	54.72946
7.3139453	7.3138404
16.366828	16.366707
28.756737	28.75666
43.930498	43.930474
46.965249	46.965237

Input data for Gauss elimination using the Program in Appendix 4

					S(M,P)							R(P)
−4	1	0	2	0	0	0	0	0	0	0		0
1	−4	1	0	2	0	0	0	0	0	0		0
0	1	−4	0	0	2	0	0	0	0	0		−100
1	0	0	−4	1	0	1	0	0	0	0		0
0	1	0	1	−4	1	0	1	0	0	0		0
0	0	1	0	1	−4	0	0	1	0	0		−100
0	0	0	1	0	0	−4	1	0	0	0		0
0	0	0	0	1	0	1	−4	1	0	0		0
0	0	0	0	0	1	0	1	−4	1	0		0
0	0	0	0	0	0	0	0	1	−4	1		−100
0	0	0	0	0	0	0	0	0	2	−4		−100

References

Allison, J. (1971). *Electronic engineering materials and devices*, McGraw-Hill, New York

Baden-Fuller, A. J. (1973). *Engineering field theory*, Pergamon, Oxford

Brown, A. (1973). *Electricity and atomic physics*, Macmillan, London

Carter, G. W. (1967). *The electromagnetic field in its engineering aspects*, 2nd Edn, Longman, London

Chikazumi, S. (1965). *Physics of magnetism*, Wiley, New York

Einstein, Albert (1968). *Relativity – the special and the general theory*, 14th Edn, Methuen, London

Hague, B. (1951). *An introduction to vector analysis*, 5th Edn, Methuen (Monograph), London

Hayt, W. H. (1967). *Engineering electromagnetics*, 2nd Edn, McGraw-Hill, New York

Jeans, J. (1925). *The mathematical theory of electricity and magnetism*, 5th Edn, Cambridge University Press

Morrish, A. H. (1965). *The physical principles of magnetism*, Wiley, New York

Putley, E. H. (1960). *The Hall effect and related phenomena*, Butterworth, London

Ramo, S., Whinney, J. R., and van Duzer, T. (1965). *Fields and waves in communication electronics*, Wiley, New York

Solymar, L. and Walsh, D. (1970). *Lectures on the electrical properties of materials*, Oxford University Press

Bibliography

The following titles may be of interest for background reading.

M. Abraham and R. Becker, *The classical theory of electricity and magnetism*, Translation of 8th German Edn, Blackie, Glasgow, 1932

S. S. Attwood, *Electric and magnetic fields*, 3rd Edn, Wiley, New York, 1949

L. V. Bewley, *Two-dimensional fields in electrical engineering*, 1st Edn, Macmillan, New York, 1948

D. R. Bland, *Solutions of Laplace's equation*, 2nd impression, Routledge and Kegan Paul, London, 1965

B. I. Bleaney and B. Bleaney, *Electricity and magnetism*, 3rd Edn, Oxford University Press, Oxford, 1983

H. Cotton, *Vector and phasor analysis of electric fields and circuits*, 1st Edn, Pitman, London, 1968

C. A. Coulson, *Electricity*, 1st Edn, Oliver and Boyd, Edinburgh, 1948

E. G. Cullwick, *The fundamentals of electromagnetism*, 3rd Edn, Cambridge University Press, 1966

K. Foster and R. Anderson, *Electromagnetic theory: Problems and solutions*, Vols. I and II, 1st Edn, Butterworth, London, 1970

W. J. Gibbs, *Conformal transformation in electrical engineering*, 1st Edn, Chapman and Hall, London, 1958

B. Hague, *Electromagnetic problems in electrical engineering*, 1st Edn, Oxford University Press, 1929

P. Hammond, *Applied electromagnetism*, 1st Edn, Pergamon, Oxford, 1971

R. F. Harrington, *Field computation by moment methods*, 1st Edn, Collier Macmillan, London, 1968

William H. Hayt, Jr., *Engineering electromagnetics*, 4th Edn, McGraw-Hill, New York, 1981

J. D. Kraus and K. R. Carver, *Electromagnetism*, 2nd Edn, McGraw-Hill, New York, 1973

P. Lorrain and D. R. Corson, *Introduction to electromagnetic fields and waves*, 2nd Edn, W. H. Freeman, San Francisco, 1970

L. Marder, *Vector fields*, George Allen and Unwin, London, 1972

J. C. Maxwell, *A treatise on electricity and magnetism*, Vols. I and II, 3rd Edn, Clarendon Press, Oxford, 1904

E. B. Moullin, *The principles of electromagnetism*, 1st Edn, Oxford University Press, 1932

L. Page and N. I. Adams, *Principles of electricity*, 3rd Edn, Van Nostrand, Princeton, New Jersey, 1958

C. R. Paul and S. A. Nasar, *Introduction to electromagnetic fields*, McGraw-Hill, New York, 1982

R. Plonsey and R. E. Collin, *Principles and applications of electromagnetic fields*, 2nd Edn, McGraw-Hill, New York, 1982

B. D. Popovic, *Introductory engineering electromagnetics*, 1st Edn, Addison-Wesley, Reading, Massachusetts, 1971

N. N. Rao, *Basic electromagnetics with applications*, 1st Edn, Prentice-Hall, Englewood Cliffs, New Jersey, 1972

J. R. Reitz, F. J. Milford and R. W. Christy, *Foundations of electromagnetic theory*, 3rd Edn, Addison-Wesley, Reading, Massachusetts, 1979

L. C. Shen and J. A. Kong, *Applied electromagnetism*, Brooks/Cole Engineering Division, Monterey, California, 1983

H. H. Skilling, *Fundamentals of electric waves*, 2nd Edn, Wiley, New York, 1948

R. V. Southwell, *Relaxation methods in engineering science*, 1st Edn, Oxford University Press, 1940

M. R. Spiegel, *Vector analysis*, 1st Edn, Schaum, New York, 1959

M. Walker, *Conjugate functions for engineers*, 1st Edn, Oxford University Press, 1933

E. Weber, *Electromagnetic fields: theory and application, Vol. 1: Mapping*, 1st Edn, Wiley, New York, 1950

Index